Tarika!

Enjoy your
world travels and
tears of the world.

Sincerely,
Barry

Silver Spoons, Mad Baboons, and Other Tales of Tea

Barry W. Cooper
International Tea Master

Silver Spoons, Mad Baboons, and Other Tales of Tea
Copyright ©2008 by Barry W. Cooper

Copyright on all photographs and maps are held by Barry W. Cooper, except for photos on pages 29 and 37, which are held by Getty Images.

Portions of chapter 25 previously appeared in *Tea & Coffee Trade Journal.*

Published by Cooper Venture Associates, Louisville, Colorado, 80027.

Visit us at www.CooperTea.com

Printed in China.

Library of Congress Cataloging-in-Publication Data available.

All rights reserved. No part of this book may be reproduced in any form or by any electronic or mechanical means, including information storage and retrieval systems, without permission in writing from the publisher, except by in the case of brief quotations embodied in critical articles and reviews.

FIRST EDITION

Cover design and interior layout by J. Charles Holt
Publishing consultation by Colleen Norwine at Sedobe, LLC
Editing by Tara Weaver
Maps by Scott Church Creative

ISBN-10: 0-9777397-1-6
ISBN-13: 978-0-9777397-1-4

*For N.F.H. (Toby) Fleming,
mentor, friend and tea man extraordinaire,
and
Stevie Finch,
who gave me shelter and set me on my life's path.*

Contents

1. Africa Is Home 1
2. A Soldier and a Spy 5
3. Safari Encounters......................... 10
4. Burning Bushes 18
5. Seventeen and Searching 26
6. Lipton Training 33
7. The Brothel 41
8. Uganda Bound........................... 47
9. Working the Fields 50
10. Ahead of His Time 56
11. Guns and Golf.......................... 60
12. Waiting on a Visa 65
13. American Ball Busters 73
14. Texas Wisdom........................... 79
15. Tea Chest Is a Piece of Shit............. 88
16. Flight Out of a War Zone 97
17. Diving into Specialty Teas 106
18. Countries of Origin: India.............. 109
19. Countries of Origin: China.............. 117
20. Countries of Origin: Formosa 127
21. Countries of Origin: Ceylon 131
22. Countries of Origin: Japan.............. 137
23. Countries of Origin: Kenya 145
24. The Challenge of Selling Specialty Teas 152
25. Herbs (Hibiscus, Rose Hips, Chamomile, Peppermint, Cinnamon) 157
26. The Secret of Blending.................. 199
27. Taking an Iced Brew to Market........... 205
28. The Celestial Years..................... 210
29. Celestial in the Rearview Mirror 221
30. Sandbar Sets Sail 225
31. Putting a Face to Cooper Tea 229
32. Summing Up 238

Africa

Chapter 1
Africa Is Home

I GREW UP IN KENYA, AS IDYLLIC, RAW, VIOLENT, AND BEAUTIFUL A PLACE AS THERE IS ON THIS EARTH. WHEN I TELL PEOPLE I AM A WHITE AFRICAN AMERICAN, A STRAINED SILENCE RESULTS. IT SOUNDS POLITICALLY INCORRECT. A WHITE AFRICAN AMERICAN? HOW COULD THAT BE? SURELY NO SUCH CREATURE EXISTS.

Well, of course we do. I suspect there are thousands of us—probably tens of thousands—scattered across America. We are the sons and daughters of the white colonies of Africa: the Southern Rhodesians; the Northern Rhodesians; the kids who grew up in Nyasaland, Tanganyika, South Africa, or Kenya. We were raised in families that spent generations on the Dark Continent, only to find ourselves homeless and stateless as Southern Rhodesia toppled into Zimbabwe, Northern Rhodesia became Zambia, Nyasaland became Malawi, and Tanganyika renamed itself to an almost-familiar Tanzania.

All of these and many more countries were the playgrounds for a pampered race of white children, raised by black nannies and devoted to sports and club life. We banded together in comfortable groups with similar likes and dislikes. Our society signed chits for everything, even groceries, so that even today I often find

Chapter 1

Here I am by the Mara River while on safari.

myself with no money in my pockets, forced to use a credit card for a two-dollar purchase. We were raised to view leadership as a natural event and mastery as a destiny. We were used to broad verandas and wide vistas of open plains and soaring skies.

We were the sons of soldiers, farmers, diplomats, missionaries, lawyers, policemen, and civil servants. We were raised with servants, gardeners, and cooks in homes that rambled on forever and were surrounded by English country gardens and huge lawns that were cut and clipped by an army of gardeners.

Our parents lived lives surrounded by chintz, politics, and cocktail parties. Their time was spent on affairs of state—and of the more pleasurable kind of affairs, as well. Their interests all coalesced into the single theme of maintaining both their lifestyle and the British flag flying over the colony, no matter what part of Africa that happened to be.

Africa Is Home

We children were raised with manners of a bygone age. We were held to behaviors and standards that, even as we were being taught them, were disappearing forever in the faraway land called England, where we would go on "home leave."

We were cultural misfits. Comfortable in our African heritage, we were at home in the wilds of Africa and attuned to where puff adders would rest, where crocodiles would slide, and how much Angostura bitters it took to make a gin pink without turning it red. We were universally raised with a Gordon's gin bottle full of water. The water had been filtered through a porous rock, and the bottle would always take a place of pride in the center of the dining room table.

At an early age we knew that rhinos could not see worth a damn but could hear you a mile away and were mean. We knew the lazy-looking buffalo with its droopy horns and droopy-looking face was by far the most dangerous animal on the plains and should never be ignored; we knew that male lions were pussy-cats and would not hurt a fly but female lions viewed us as a meal. We knew to shake out our slippers in the morning to avoid crushing centipedes against our toes and that earwigs could crawl into your ear if you slept on an infested pillow. We knew that Oxford and Cambridge were the only universities in the world and that God was an Englishman.

As boys we were taught the manly arts of rugby and cricket and boxing. We viewed girls as desirable objects—remote, unattainable, and to be protected with your life if necessary. We knew that anyone suggesting something different was a liar. "Manners maketh the man" was not just a saying to us, it was the law. We were totally safe and secure in an entirely alien land.

We were as much at home in Kenya as the smallest *toto* (child) running around naked in a village, and the same love of Africa still

Chapter 1

runs in our blood. We matched in our uniforms at our English schools—short shorts; striped ties; and blue, red, or green blazers. We were divided into houses at school, all bearing the names of English, Scottish, Welsh, and occasionally Irish noblemen. The names Clive, Wellington, Hawk, Frobisher, and Rhodes were familiar to us before we even opened up our history books.

We regarded these men as heroes. Clive of India was a lone adventurer who conquered one of the most populated lands in the world, making it the jewel in the crown. We were being raised for that same destiny. I think we must have been interchangeable at that time. You could have whisked a child out of school in Salisbury, Southern Rhodesia, into a school in Nairobi, Kenya, and he would not be noticed unless his blazer was of a different color.

We were raised to recognize a gentleman by his punctuality, the force of his personality, the tone of his voice, and the bearing of his shoulders. Young district commissioners would arrive from London to take on responsibilities for administrating swathes of territory the size of southern England. No one questioned their ability or right to do so.

As historians delve into the past and write about the policies and attitudes of the times I grew up in, I see the arrogance and paternalism that was not apparent to me as a child. But I cannot quibble with the sheer joy of my childhood in Africa; I was a child and saw things as a child does, with no political bent whatsoever.

It was a doomed society even as we grew up, and our fate was to be the immigrants of a new age. We were a generation of children who were destined to be travelers, to be the storytellers of a life that was disappearing as we were living it. We were to take with us the memories and experiences of growing up in Africa—as white children, but as native as could be—and go out into the wide world to tell our story.

Africa Is Home

After Independence we scattered like seeds in the wind. The children of the settlers had no place to settle. Kenya was my childhood home, and England is my birthplace; America is now my homeland.

I have watched in horror as Africa has disintegrated into a morass of warlords, warfare, corruption, incompetence, malpractice, and illness. Despite all the faults that its inhabitants bring to it, the continent remains serene and magnificent, soaring above the petty vices of its rulers. It retains its ancient soul. Africa gets in your blood; once you have experienced the inner pull of the Dark Continent, life can never be the same again. It tugs at your heart and brings you back.

Africa was the birthplace of mankind. It is terribly old, and its recent history of fresh new nations also makes it terribly young. Africa gave me a fractured cultural background but one that has made me comfortable in any culture.

Chapter 2
A Soldier and a Spy

I FIRST SAW A TEA BUSH IN MALAYSIA WHEN I WAS FIVE YEARS OLD. IT WAS 1949. MY FATHER, A PROFESSIONAL SOLDIER, WAS ATTACHED TO THE 7TH BATTALION OF THE GURKHA, TRAINING TROOPS ON COUNTERINSURGENCY TECHNIQUES. MY MOTHER WORKED FOR MI6, THE SECRET INTELLIGENCE SERVICE, AND WAS RESPONSIBLE FOR PAYING THE INFORMERS WHO WERE USED TO HELP TRACK DOWN THE COMMUNIST TERROR GANGS. WE WERE ON OUR WAY TO THE BRITISH ARMY REST CAMP FOR A HOLIDAY.

The British Army rest camp was in the Cameron Highlands, a verdant patch of mountaintop that was cool and misty and surrounded by tea bushes. Unfortunately, it was also surrounded by Communist guerillas. To get to the rest camp, you had to join a scheduled convoy. My father did not like convoys. "Why tell your enemy what time you are going to leave and then drive slowly past him?" I heard him tell my mother. Instead we traveled with a soldier who had a machine gun in the back of the car. The car was a black Hillman Minx. I do not think there were too many private cars in Malaya at that time, but my parents had one.

My father was driving, my mother was in the front seat, and my sister and I were in the back, along with a sergeant who had a machine gun in his hands to defend us in case of an ambush. As I look back, I now think he was there to make sure my mother and father were not captured. Killed if necessary, but not captured.

We saw a Communist guerilla on the way up the mountain. He was sitting on a mile marker, a rifle in his lap. We were past him and round the corner before he moved. My father accelerated the car, though.

As we climbed up toward the official army rest camp perched atop the hills, we drove through rolling hills and valleys full of these little bushes. I remember looking out and wondering why all the bushes were the same size and how all the rows could be equidistant, and I was fascinated by the way they followed the contours of the land.

Of course, I now know that contoured planting prevents runoff from the rain and that each tea bush is the same height to allow pluckers to harvest the tea leaves. But to a five-year-old boy, those orderly, precise rows were a source of wonder and amazement.

My father had spent the Second World War training Special Operations Executive (SOE) agents in counterterrorism tactics. Winston Churchill created the SOE in July of 1940 to organize sabotage and general mayhem in occupied Europe. Father worked at London's Hendon Police College, teaching agents how to blow things up, dismantle railroad tracks, kill people with a single punch, and other useful day-to-day needs for an undercover agent.

He trained Germans who wanted to fight Hitler and British and American prisoners who were given the option of a severe prison sentence or a short life as a commando. If they were killed, they received the honor of having died in action. He told me of all manner of men who went silently into the night, back to Europe

Chapter 2

My family: Roland, Susan, Sylvia, and myself.

to fight. My father lost both his brothers in the war, and like most hard men of a hard profession, he was very gentle around his family and friends.

When the war ended, Communists infiltrated the Malaysian Peninsula, and the British Army was sent out to repel the guerillas; my father was sent to train them. That is how I ended up in Malaya and also how I found myself in Kenya five years later for my second encounter with the tea bush.

The Mau Mau Uprising, a revolt in Kenya against British rule, began in the early 1950s, and by 1954 Kenya was a dangerous place to live. Cattle were being hamstrung, people were being killed in horrendous fashion, and a pall was being cast over a beautiful country. Our family arrived in Kenya in March of 1954. My sister and I did not feel the effects of the emergency, although we were firmly told that we had to be in the house at dusk. I slept in a room with a loft that had a steep ladder; if I heard noises or gunshots, my instructions were to take my sister, climb the ladder, and pull it up behind us. We were then to sit and wait for our parents. In the loft were switches to battery-operated sirens. I was to activate the sirens so that the neighbors would hear that we were in distress or being attacked.

I caught my second glimpse of tea bushes as we took a visit to the Limuru Brackenhurst Hotel, perched in the highlands of Kenya about fifteen miles outside Nairobi. I recognized the bushes immediately. Now it was certain. Gazing wide-eyed at the rolling hills, I felt a distant call that came to define my life. This fascination only intensified when my family settled in Africa.

Chapter 3
Safari Encounters

In Kenya, we settled in the heart of Kikuyu tribal lands. Our home was surrounded by coffee bushes on one side and tea bushes on the other. Five miles away, the deep gash of the Great Rift Valley soared downwards, and I was later to spend long expanses of my time roaming around the flatlands of the Rift.

Each year, just before the rains were due, the Maasai set fire to the Ngong hills on the edge of the Rift. No one knew how the Maasai could tell that the rains were coming, but as soon as the hills were seen enveloped in smoke, everybody knew the rains would start.

After the burn, all that was left was scorched soil and sooty rocks. Everything was burnt back, but with the rains the grass would return, thick and lush, and the Maasai would bring their cattle to graze. It was part of the ancient rhythm of their lives.

On the far side of the hills there was an escarpment that was full of game. Lions, giraffes, zebras, Tommy gazelles, dik-diks, hyenas, tick birds, and nesting swallows all lived in profusion.

The escarpment was not part of the Nairobi Game Park—that tame tourist attraction was just outside the city. Nairobi was called the city where you could hear the roar of a lion. Maybe it had been that way before, but that time was long gone. Now you could drive up the road toward the Ngong hills and turn left for the Wilson Airdrome or turn right into the game park. It was like going to the zoo—very civilized.

The escarpment was not like that. It was wild, well away from roads and from people. To go there was to enter a dangerous world where you were the weakest to walk the surface—unless you had a gun, and in those days, guns were not allowed. The Mau Mau wanted guns, so to walk around with one was to invite trouble. The week that we arrived in Kenya, two schoolboys had been murdered for their air rifle.

> ## LOCAL LORE
>
> Hippos are considered to be one of the most dangerous animals in Africa. They kill more people every year than all the other African animals put together. In addition to them always being a threat to tip over a canoe, hippos come ashore at night, and anyone who has the misfortune to get between them and the water is likely to be attacked. With a land speed approaching thirty miles an hour and twenty-inch incisors, these "aquatic teddy bears" are anything but cute.

What kept you alive in the escarpment was knowledge. You had to know that a lion on a rock surrounded by females was good. A male lion alone on a rock was bad; that meant the females were hunting. You needed to know never to walk behind a giraffe; its rear feet would rip you from tip to toe. The most dangerous animal by far was the hyena. The hyena is clever, ruthless, and powerful; even the lion is afraid of the hyena and its crazy laughter. You learned that snakes liked shadowy rocks and scorpions liked sandy crevices. Rhinos could not see, but they could hear and smell, and they were bad tempered.

Chapter 3

The Maasai would walk in the escarpment and be a part of it. Their ochre blankets and colored beads were as much a part of the scenery as the gray hide of a rhino. The blankets, under which they wore nothing, would flap around in the breeze; in Nairobi, tourists would ask to take pictures and then wait for a gust of wind. The moran, or warriors, would stand there, proudly arrogant in the pose and smiling at the stupidity of these white people who wanted a picture of their private parts.

The escarpment was too far away from my home to be my backyard. To go there took planning. But to me its flat-topped thorn trees, endless vista stretching away to the Magadi salt flats, dry heat, and remoteness were the real Africa. I would go there with friends during the school holidays to play in the bush, and it would always scare us with its simple wildness. It would have terrified my parents if they had known we were out there. But they never asked.

When I was ten, my grandparents came over from England to see the real Africa. We drove up to the Queen Elizabeth National Park in Uganda. It was a wild place, like the escarpment.

On the way to the lodge, we came across a herd of elephants and stopped to watch them cross the road. My grandfather opened the car door, got out, and walked toward them. A large cow separated from the group and challenged him, her head lowered, her ears raised in warning.

My mother yelled at my grandfather to get back in the car, and my grandmother fainted. My father got out of the car to bring him back. When my father got out of the car, my mother's voice went up ten octaves and took on a hysterical note. My sister started to cry, and I watched Dad as he slowly talked Granddad away from the matriarch.

Granddad must have realized that his attempt to get a photograph standing alongside the herd had been an error in judgment. He was now trying to extricate himself by waving his apology to the cow and backing away. This friendly apology by Granddad was misunderstood by the cow, who raised her trunk and screamed. Anyone who has lived in Africa will tell you that this is the last warning before an elephant charges. My father knew it, I knew it, and my grandfather hadn't a clue.

My mother did, however, and she clamped her hand over the horn. The sound drew the cow's attention away from Granddad and to the car. The cow started to charge just as my grandmother was coming around from her faint. She yelped and fainted away again. My mom slid across into the driver's seat, slammed the car into reverse, and backed down the road.

When the cow saw the car backing away, she stopped charging and looked for the other distraction, but my father had pulled my grandfather away from center stage and all was well. The cow went back to the herd she had protected, and my grandfather went back to the wrath of his wife and daughter. After that we took great care to watch out for Granddad, who looked upon Africa as a large zoo.

To go on safari in those early days was a true adventure, and the road safari we took with my grandparents was one way of seeing the country. We drove from lodge to lodge and took side trips to see the animals. The other type of safari was the tented trip, where you set up a camp and lived there for weeks. We did that, too. That took us into the Congo, and we crossed Lake Albert in dugout wooden canoes, my grandfather singing the Eton boating song. The lake was full of hippos, who look friendly but are not. They like to come up beneath canoes and tip them over. I knew it, my father knew it, and my grandfather didn't have a clue.

Chapter 3

Percy Byers was our safari guide. He worked for the railroad—to be more specific, the East African Railways and Harbours Corporation. He had a collection of gleaming rifles; their shiny wooden stocks and glistening oily barrels appeared ominous in the gloom of his study. He was a true white hunter, not the type who took the tourists out and wore a jacket for dinner, but a real hunter. Percy had to go out with the railway crews to survey the lines and repair the tracks. He shot game for food and to protect the workers. He never shot for fun.

Percy was stocky and bandy-legged. He wore a big, floppy hat with a strip of zebra skin around the crown. His eyes would crinkle as he spoke. His wife, a small woman in flowered dresses who knitted, traveled with him. She would sit in a canvas fold-out safari chair and knit endless reams of woolen stuff. I never saw a complete garment of hers.

The Mara River was wild, wilder than the escarpment. It was deep bush—you depended upon your own resources in the Mara. To get there, you left Nairobi on the Nakuru road, and anywhere between mile forty-five and mile ninety, you turned left off the road and headed southwest until you hit the river. When you reached the river, you made camp because you had arrived.

The first order of business was to chop down a tree and lay it against another tree to make a funk ladder. This gave you somewhere to run if you were attacked by something large. After that you would set the perimeter, assemble the tents, set up the kitchen tents and the cook's quarters, and then go and find the game.

It was not difficult. They were right outside the perimeter. Within the space of an hour, the trees around the tent would be alive with baboons and monkeys chattering their news. This gossip was heard by the hyenas, and those animals, with their distinctive sloped shoulders, would soon be seen skulking among the trees.

Safari Encounters

Olive baboons. It's not their teeth that makes them dangerous—it's their intelligence.

Then the birds would arrive, looking for leavings, and soon the camp would become a cacophony of sound—chirping, chattering, gibbering animals, all watching every move you made.

The baboons were the worst. The females would sport painful-looking backsides of violent red. They strutted on all fours with their arses jutting up in the air, splashes of truly impressive color paraded before the world. They had three-inch fangs, large snouts, and small, piggy eyes. Baboons were not to be messed with. No one had to tell you this. One look at their teeth (or tushes) would tell you this was an angry animal to be avoided.

We had to move camp once because I threw a rock at a baboon. I had gone down to the river to fetch water. On the way down to the Mara, I came across a couple of baboons sitting under a tree. They were too near the path for my comfort, so I made them move by throwing a rock in their direction. They looked offended but got up and loped away.

Chapter 3

I walked down to the river, being careful to avoid the croc slides. These were greasy mud slicks that a crocodile made with its tail where the bank fell steeply to the water's edge. The crocodile would then clamber on its short, stubby legs up to the top of the slide and wait for a small buck or a person to walk to the river. Then, in one convulsive surge, it would slide down the bank, whipping its tail around the unsuspecting victim, who would then tumble into the water to be drowned. The carcass would be stuffed under the riverbank until the flesh rotted. The croc then returned to nibble on the tenderized, bloated body at its leisure. It was a fate to be avoided.

Finding a flat rock away from the croc slides, I filled up the canvas buckets and started back. As I rounded the bend, the two baboons I had pelted with a rock met me. They had come back with about thirty of their friends. I dropped the buckets and picked up a rock. They all reached down and picked up rocks. I dropped my rock. They didn't drop theirs. I started to run; they started to follow. I started to run really hard, and I didn't look back. I made it back to the camp, closely followed by the pack. I flew into my father's tent, out of breath and begging forgiveness.

It was not immediately clear why I was begging for forgiveness. It soon became so. Every time I appeared, the baboons would leap up and down and throw broken branches at me. They formed a perimeter guard that would alert the pack whenever anyone tried to go to the river. Then they would all leap from the trees and throw rocks at the water bearer. After two days of this, we broke camp and moved five miles downriver. I had learnt my lesson: Never throw rocks at a baboon.

Years later when I brought my youngest son Rory to Africa to be christened, we stayed at the Muthaiga Club in Nairobi. A baboon watched us every day. On day three, it came to the door of our bungalow and tried to steal my son. The kidnapping was thwarted by my eldest son Brad, who was seven at the time. I wondered if the baboon tribe had put the word out that I was back in town. Impossible, of course, but who knows—this is Africa.

Chapter 4
Burning Bushes

BY THE TIME THE MAU MAU REBELLION HAD SUBSIDED TO A FEW JUNGLE FIGHTERS LEFT TO THEIR OWN DEVICES DEEP IN THE FORESTS OF MOUNT KENYA, MY FATHER HAD DECIDED TO RESIGN FROM THE ARMY AND TAKE UP CIVILIAN LIFE.

We remained in our home, surrounded by the coffee and tea bushes. I entered the lost years of teenage-hood, when banishment to my room became the most common form of parental discipline. This punishment taught me the joy and the endless pleasure of reading alone. By age fourteen, I had finished *War and Peace* and had found Tolstoy's notes by far the most interesting part of the book. I read *Seven Pillars of Wisdom* by T. E. Lawrence and formed an opinion of the man that no movie or subsequent book has ever changed. I also read all of Winston Churchill's volumes on the Second World War, Eisenhower's *Crusade in Europe*, Douglas Bader's *Reach for the Sky*, P. R. Reid's *Escape from Colditz*, and every other escape story I could find. I devoured books. I loved novels by Georgette Heyer and other mid-seventeenth- and eighteenth-century romances.

Reading was a passion, an open door that took me to other worlds. I experienced burning deserts with Lawrence and

Napoleon's battlefields with Tolstoy. I soared with fighter pilots in the battle of Britain and quivered through depth charge attacks on British submarines. I suffered on the road to Mandalay, and I fought at Singapore. I was never bound by the fence posts of our Kiambu home. I was a reader. I was a time traveler. True punishment was to be given a task.

"I want you to clear the hill behind the boys' quarters of lantana, Barry." My father's voice was firm.

"Clear the hill? You mean the whole hill?"

"Yes, the whole hill."

"But that will take forever!"

"I know." There was a streak of cruelty in my father.

I faced the hill, a reluctant Jerroggee at my side. Jerroggee was my playmate and my partner in all things. He was fifty years old, with the heart of a child and the wisdom of a teenager. His role in the household was general helper, but mainly he used to hang out with me when I was around. He had no front teeth. He explained to me that his mother had knocked them out so she could feed him if he ever got lockjaw. The lack of teeth sometimes made him difficult to understand, but his grin was infectious. Even my father would smile when Jerroggee would recount an exploit.

We faced the hill together. Lantana was vicious, a tough, flexible bush that had to be hacked to death. The wands of its branches were too thin to withstand the blow of a *panga*, the curved African blade that we used to cut wood and clear brush. The branches would slide away and then whip back at you. Curved thorns covered the branches. They would slice into the skin, leaving a thin trail of blood.

The broadleaf lantana shed its foliage constantly, so the bushes grew in a carpet of thick, dried leaves. This was home to snakes,

Chapter 4

LOCAL LORE

The Maasai are a semi-nomadic people living in Kenya and Tanzania. Numbering some five hundred thousand people, they are one of the better-known African indigenous tribes.

The Maasai believe that all of the cattle on the planet are a gift to them from God (*Enkai*), and their frequent cattle raids on neighboring villages aren't to steal cattle but merely to take back what is already theirs.

A Maasai traditionally could not take a wife until after he had killed a lion, a ceremony that was done in groups using spears and shields. The boy who speared the lion first was revered and afforded special privileges throughout his life. This is one of many ceremonies that is little practiced now due to changes in the political and physical landscape.

rats, and hump-kneed spiders that would scurry over your shoes as you disturbed their habitat.

Lantana would grow to six to eight feet in height, so you worked in a constant gloom. As you worked, a fine cloud of moldy spores would rise up from the leaves, and a shower of insects would fall on your head from the branches above.

It was garden hell.

I was prepared. On my head I wore a soft, broad-brimmed safari hat to stop the insects from falling down the back of my neck. I donned a long-sleeved, buttoned shirt and safari trousers tucked into my boots. I topped all this off with a pair of leather gloves and sunglasses in case the wands sprang back into my eyes.

Jerroggee wore his normal *ganza*, a length of cloth tied up between his legs. He was shoeless and hatless, but he was ready. Together we entered the lantana, sharpened *pangas* at our sides.

Hacking and slashing, we worked our way into the hill. The only way to effectively cut lantana was to grasp a branch, bend down to cut it at the base, pull it free, and throw it to the ground. After half an hour, I was dripping with sweat, my arms were sore, my hat was covered with crawling insects, and my shirtsleeves were snagged in a hundred places.

Jerroggee was battered as well. His legs bore testament to the thorns. We stopped. We had cleared about twenty feet square. The hill loomed before us, at least five hundred yards wide and forty yards deep. It looked awe-inspiring.

"Let's take a break, Jerroggee," I said. I could tell that this was the best suggestion he had heard all morning. We sat on a log and looked up at the hill.

"The Maasai are burning the Ngong," Jerroggee said. I looked over and saw the telltale spirals of smoke.

"The rains are coming," I said. Jerroggee nodded, his lips pursed in thought. At that moment a large snake slithered around the freshly cut stumps of the lantana and disappeared up the hill into the thick leaf cover. Jerroggee pursed his lips even more and drew his breath in sharply.

We continued to sit on the log.

"Perhaps we can learn from the Maasai, Bwana Barry," he said. I looked at him. He nodded toward the hill. "We can use fire to clear the old leaves. Then it will be easier to cut."

I looked sideways at Jerroggee. Then I looked up at the acres of lantana. I thought of the snake we had just seen disappear into the undergrowth. "Good idea," I said. "I know where there is a can of petrol."

Chapter 4

"We will just make it a small fire, Bwana Barry. Just something so we can burn away the *taka taka*."

It was the work of moments to get the petrol, pour a thin stream across the roots of three lantana bushes, and strike a match. It worked! Soon a nice little skirmish line of fire was working its way up the hill. The thick undergrowth developed a satisfying reddish glow as it simmered. The leaves of the lantana curled back in pain as the fire reached out and singed them. The whippy branches began to blacken. I felt an emotional satisfaction that I was causing this vicious plant agony. *So there!* I thought.

Jerroggee and I carefully advanced through the smoking undergrowth and began hacking at the blackened wands. From nowhere, the lightest zephyr of a breeze appeared—just a kiss upon the cheek. I looked up. The skirmish line had felt it as well. The tendrils of fire wavered, then leapt up to six feet in height. We heard crackling sounds as the top branches of the wands met the flames. Blackened leaves began to waltz into the sky.

"Aaaaahhhhhh," I heard Jerroggee say. It was a sound that was familiar to me. It was not good.

"Get the water hose, Jerroggee." I needn't have bothered. Jerroggee was already galloping across the lawn in the direction of the garden boy. He soon reappeared with the hose and the garden boy, who stood there grinning.

I yelled "*Maji!*" (water). I had to shout because the crackle had developed into a roaring sound. I looked back at the fire. The skirmish line was no more. In its place was a wall of flame, flickering and consuming all in its path. Birds were shooting into the air, and a plume of dark smoke covered the hillside. I began to cough in the smoke. The heat was intense.

"Ahhhhhhhhhhhhhhhhhhhhh!" Jerroggee was in fine form today. The garden boy started to snicker.

"Bwana Barry." I looked at Jerroggee standing with the hose, which was spluttering a supply of water that was completely inadequate for our needs. "Bwana Barry, Bwana McCloud."

I jumped in my skin. Our next-door neighbors' house was on the other side of the hill. It was a good half-mile away, but the way the flames were surging up the hill, it would not be long before they would reach the crest. On the other side of the crest were a series of vegetable gardens, the main house, the servants' quarters, and then row upon row of seedling tea bushes—all waiting in the path of the roaring inferno I had created. Actually, it had been Jerroggee's suggestion, but legal parsing was not going to save me if I burnt down John's place.

John McCloud had a tea nursery. Traditionally, tea propagates through seeds. When the seed germinates—when it cracks its shell and lets a root wiggle out—it will take a year before the slender plant has a chance of survival. The roots love rich, acidic soil and do not flourish in dry environments. The seed develops a taproot that seeks water and can grow up to five feet long—longer if it is exposed on a bank.

Within five years, the slender bush develops enough leaves to begin contributing to the overall yield of the estate. Every seven years, the bush is pruned so that its top, or table, remains flat and easy to pluck. It is the fresh young buds that are picked, and they are always, or almost always, surrounded by two little leaves—hence the expression "two leaves and a bud."

Nowadays, no one expands by using seeds. Today, tea bushes are cloned by taking cuttings from high-yielding bushes and then growing hundreds of such cuttings so that the tea fields can be calculated to produce fairly precise amounts of tea. This was John's setup. He had rows of cutting sleeves, then beyond them yearly cuttings in beds, and beyond them the plants ready for sale.

Chapter 4

A roaring inferno would devastate the garden, but at that point it would have already consumed the house, and I would be a dead man.

"I'll tell him," I yelled and ran down the driveway and then over to McCloud's home. I remember flying over the ground, but I arrived at his front door exhausted. I pounded on the door with no sense of decorum left. When Mr. McCloud opened the door, all I could do was point at the forest fire that was now running the entire length of the ridge.

"Jesus Christ" was all John said before disappearing back into his house. He gathered his house servants to fight the fire. I sat on the doorstep, my world in tatters. There was no way of hiding a blackened hillside from my father. I thought about running away to sea. Did they still hire cabin boys?

John McCloud had cleared his land of lantana many years prior. The fire had nothing to feed on and slowly died away. It managed to consume about an acre of his maize field before slowly ending its life by the irrigation ditch.

As John walked back toward his house, I saw my father's car pull into the driveway. It crunched its way up to us. My father got out and looked at the blackened hillside, the burnt maize field, then at John, and finally at me.

"I saw the smoke, John. Thought I would come back and see what my son was up to. Sorry about the maize. May I settle up with you later?"

"Of course, Roland. Gave us quite a scare. Glad Barry warned me in time."

My father looked at me. "Kind of him," he said in his driest tone.

I thought of various explanations, but none came to mind apart from the truth. "The Maasai burn the hills. I thought it might work here."

"The Maasai burn unpopulated, remote grazing areas, where the grass is less than two feet high, Barry. They do not burn lantana. No one burns lantana. If you try and burn lantana, it tends to catch fire. It spreads. Only an idiot would try and clear lantana by burning it. Where's Jerroggee?"

But Jerroggee was nowhere to be found. He had a knack for disappearing. It was one thing he never taught me. I was confined to barracks again; I don't think my father knew what else to do with me. I was forbidden to touch anything or do anything. I was just to be.

Chapter 5
Seventeen and Searching

Though I left high school with surprisingly good grades, I wished to start my life and not continue the slog of enforced education.

I had a clash of wills with my parents, and I won—mainly because my parents were so involved in fighting each other, they had no time to include me in the battles. They gave up, and I was free to enter the workforce. I wanted to write.

Through my father's contacts, I was given an opportunity to join Nairobi's local newspaper, the *Daily Nation*, as a junior sports reporter. Within two months I had my own column in the Sunday edition, writing about the comings and goings of teenagers in and around Nairobi.

In those days, the city was the center for all the overseas correspondents covering central Africa for English newspapers and magazines. They were a group of hard-drinking, hard-living, worldly wise men, and here was I, a seventeen-year-old mascot.

It was 1962, and the Belgian Congo was going up in flames. The Simba, a group of drug-crazed rebels, were shooting and raping their way across Kinshasa. Dag Hammarskjöld, the UN secretary-general, was shot out of the sky over the Congo. Refugees were flooding into Uganda and Kenya.

The correspondents were in the midst of a period of turmoil, danger, global politics, confusion, and great stories, and I was in the midst of the correspondents. It shakes me now to think of the risks I ran in those days, but fueled by Martel brandy, the bravado of being one of the guys, and the rough-and-ready camaraderie that is part of being in an elite group, I lived a life way beyond my years.

At this point, my parents had given up the ghost of their marriage, and I was living with my father. He saw where my life was heading and took me aside, gave me a round-trip ticket to England, and told me not to use the return half.

I was on my way. My hard-drinking friends gave me a farewell party and a letter of introduction to the Newcastle *Evening Chronicle*. It had come my turn to go, though I didn't have a clue what was waiting for me. I was only seventeen.

England was a shock to my ego and my system. I was cold and I was anonymous, one of millions. I took jobs as a construction worker, a demolition worker, and a garage mechanic as I waited for the golden invitation to arrive from Newcastle. When it did arrive, I took the next train up to that grim city.

Newcastle nestles against the North Sea. It is opposite Norway, and it is cold and wet most of the time. Its first major industry was coal mining; it made a lot of sense to me that they would want to spend as much time as possible underground.

The editor read my cuttings and offered me a position. The newspaper had a scholarship program; I would work at the newspaper and go to the local university. The world lay before me. I lived in Whitley Bay, a seaside resort that was stark and depressing. My landlady was Mrs. Leach, a large-bosomed lady who cooked very greasy eggs for breakfast.

Chapter 5

The horrible reality of Newcastle slowly settled upon me. I could not understand the local people when they spoke; the thick Geordie accent was beyond me. This was a major disadvantage for a reporter. After the fourth or fifth "excuse me, can you repeat that," the interview trails away. It was a problem I could never overcome.

I was doomed from the start. A work routine that took away every weekend—so that when I had a day off, everybody else was at work—did not help. The city was old and broken down, its industries of coal and shipbuilding a thing of the past, about half the population was unemployed, and I could not understand the people. I decided to make a clean break. I was not destined to be a reporter, certainly not in Newcastle.

I moved to London. I arrived penniless. Years of tradition pulled me to the security of the armed forces, so I enlisted in the Royal Marines. A full day of examinations—medical, physical, and mental—ended with the surprise news that they would contact me in a month. They could not take me then.

So I got a job in a so-called breakers yard, breaking up cars. They gave me an oxyacetylene torch and led me to a wrecked car at the back of the yard. The foreman, Fred, showed me how to work the torch and then told me to cut the car up.

A therapist could not have chosen a more healing task. I set to work, and I contentedly reduced the car to bits and pieces. They called me "his lordship" because my accent was different. There was Fred the foreman and Mr. Blackman the crane operator, an old hunched man with a spotted scalp and very little hair. His son, Joe, drove the pickup truck to collect all the wrecks, and the owner of the yard, Mr. Smith, wore an elegant topcoat.

As the marines processed my application, I thought about where life was taking me. My military ancestry included relatives who

Seventeen and Searching

The breakers yard was very cathartic, as I spent the day hacking up cars.

had signed up under the same circumstances and made the military their home. Did I want to fight foreign wars in far-off lands? My father, his two brothers, my grandfather, my great-grandfather, and my great-great-grandfather had all joined the colors.

I decided to choose a different path. I chose not to go. When the sign-up papers arrived, I sent them back. Wherever my life was to lead me, it was not to be down the throat of an enemy musket.

I remained in the breakers yard for two years. My muscles became hard again; my hands became callused. Eventually, I bought jeans, real workman shirts of blue denim, and work boots. I learnt to strip an engine of any part for resale. Pistons, carburetors, driveshafts, rear axles—you name it, I could remove it.

The crashes that came in were spectacular. The yard must have been full of ghosts at night. I have no way of knowing how long I would have stayed there. One morning I woke in drastic pain.

Chapter 5

> ### Local Lore
>
> In the early 1960s, a group of Congolese rebelled against their government. Being animistic people, the Congolese were told by their witch doctors that they were immune to bullets and would turn into Simbas (Swahili for "lions") when entering battle. In a sense, the magic worked—forty rebels and their witch doctors stormed the city of Stanleyville and routed a military force of over fifteen hundred men who were armed with mortars and armored vehicles. The Simba never fired a single shot. Numerous other fights had similar outcomes, with hundreds of well-armed militia men fleeing before the Simba in terror. The rebellion only lasted a year or so, but during that time, the Simbas killed around ten thousand civilians. I'm not sure how that was much of an improvement over the previous government.

The doctor made a house call; he diagnosed gastroenteritis. The pain would not go away, and my only comfort was soaking in a bath of warm water. I began to vomit up green bile, and twenty-four hours later the doctor said I might have appendicitis and sent me to the hospital in an ambulance. After twelve hours and a lot of tests, they found I had a stone lodged in my urethra. They told me that they were very surprised to find a twenty-year-old with a stone in his kidney, but after forty-eight hours of pain, I was so grateful for a diagnosis that I didn't care. The pain was so bad I had taken to clutching a Bible. They operated on me and pushed the stone back into my kidney. I spent a week waiting to see if it passed. It didn't, so they cut it out.

During the week I was waiting to pass the stone, I served tea to the other patients in the ward. I saw men come in; I would give them tea. The next day the bed would be empty, stripped of its mattress. The patient had died in the night.

After the operation, I was in the hospital for a week recovering. They sent me home with a scar that runs from my backbone to my groin and instructions to not do any heavy lifting for six months. My career in the breakers yard was over, and I began to cast about for other options.

Angus, my sister's boyfriend and son of the manager of the Mau Forest tea estate in Kenya, was working for Lipton as a trainee tea taster. He was posted to Calcutta. A position was open, and Angus suggested I write to Lipton. I did, and after a series of interviews, I was asked back for a final meeting.

The tea director at Lipton, a man named Wilmshurst, practically sat on the right hand of God. Tall and imposing, with an eye that could freeze a head waiter or a poor trainee at fifty paces, he was not one to trifle with. Our final meeting went well. Not a word was spoken of tea. We discussed cricket and the wisdom of playing a man at short leg (not a good idea, we both agreed, unless the bowler was deadly accurate). I felt I made points with that. We both loved rugby, so I thought things were going well. When he stood up, I was sure the interview was over. Instead, he asked me if I knew the chairman.

"No." I had never met the man.

"Well, he appears to know you," was the response. "And he would like to see you."

Stunned, I followed Wilmshurst down a long corridor into a huge office with a huge desk, behind which sat a smiling Buddha of a man.

"Here's Cooper," said Wilmshurst, and he left.

Chapter 5

"Sit down my boy, sit down," said the Buddha. "Now tell me, I see that you have given Gilbert Lee as a reference. Do you know Gilbert well?"

"He's my uncle, sir," I replied. He was also chairman of all the British Airways associated companies, hotel chains, food companies, and assorted airlines around the world. I thought he would be a good person to put down as a reference.

"Ahhh," said the Buddha. "So Kathleen is your aunt?"

In a flash of light I saw where this was going. "Yes, sir. Actually, Kathleen is my father's sister. She really is my aunt."

"Ahhh. How is she, my boy?"

"She is as stunning as ever, sir," I said, which was true. My aunt had movie-star good looks and long slender legs. Even at fifty, she was a knockout.

"Ahhhhhhhhhh," was the only response for a moment. A long pause followed. "I squired your aunt to many functions when I was a younger man, back in the '30s," the Buddha said with a whimsical smile. "Do give her my best when you next speak to her."

I was in. My career in tea had begun.

Chapter 6
Lipton Training

I HAD COME HOME. AT THE FIRST SIGHT OF A TEAROOM WITH ITS WHITE PORCELAIN CHINA, I KNEW THIS WAS WHAT I WANTED TO DO. TASTING TEA CAME EASY FOR ME. THE MILLIONS OF LITTLE FACTS AND DETAILS THAT YOU HAD TO LEARN WERE SALVE TO MY SOUL.

I absorbed tea. I loved its heritage, the stories, the smells, and the tastes. I loved the tradition and the feel of the different teas. It has never changed. To this day, I am happiest tasting and blending tea.

The tea buying and blending center for Lipton was at Allied Suppliers, an aged building in the heart of London Cockney land. A single elevator that could hold two people ran up to the top floor. From there, you had to walk through the warehouse and past the men opening tea chests with claw hammers, pulling out the nails that fastened the metal bands to the plywood. Then you went on past the blend lines, where the tea chests, originating from around the world, were lined up in blend sequence to be toppled into the huge blending drum. The drum groaned as it slowly revolved, tipping the tea endlessly until it was completely blended. Then finally, at the far end of the building where the northern light poured in through huge skylights, you entered the hallowed halls of the tasting room.

Chapter 6

> ### Tea Tales
>
> All tea comes from the same species of tea plant, called *Camellia sinensis*. It is the process that makes a tea different, not the tea bush itself. Once the leaves are picked, they are allowed to oxidize for different periods of time. This produces different colors in the leaves and also different flavors when the tea is steeped. White teas are oxidized the least, followed by green, oolong, and black. Black tea is the most popular tea in Western countries, and it usually has an oxidation period anywhere from two to three hours.

I was told that tasters always needed northern light, as it was constant and we would not suffer from bright sun or fading shadows that might influence our judgment. Such an understanding ensured that tea tasters got the best view a building had to offer.

The tasting room was divided into multiple counters, long benches on which the tea was prepared. To the right as you entered was the blending counter, ruled over by Mr. Alford, a huge man with a mighty girth and a friendly manner. He prepared all the blends for the Lipton range of products.

Next came the Ceylon counter, which brewed up only Ceylon teas, and on the far left was the Assam, India, and "other" counter, which brewed up everything else from Vietnams to Turkish and all the way through to the finest Darjeelings and Assams.

The entire tasting room was surrounded by hundreds upon hundreds of shelves filled with heavily lidded tin boxes holding last week's auction purchases. The room was a hive of quiet industry: a steady hum of kettles spouting steam, lids clattering as teas brewed, master tasters and their apprentices slurping and spitting, and all permeated with the wonderful smell of freshly brewed tea. It was heaven!

I started out on the Ceylon counter. A young Ceylonese taught me how to weigh and brew the teas. Within a week, I was competent and fast at preparing the hundreds of cups that we had to taste each day.

We would heat up large copper kettles with long, curved spouts on a gas stove. As the water was heating, we measured two pennyweight (three grams) into a handheld balance and tipped the loose tea into a white porcelain mug with serrated edges. We then took the loose tea and went down the long counter, which held six trays with eight mugs and bowls each for a total of forty-eight sets. When the water boiled, we ran down and poured an inch of boiling water into the bottom of each mug so that each started to brew at the same time.

Then we returned and topped off the mugs and placed the lids on. When we had gone the length of the counter, we set the timer for six minutes. When the timer sounded, we tipped the mug into the bowl so that it was at an angle, and the tea poured into the bowl and the leaves were trapped against the serrated edges. Then we went back and tipped the tea leaves into the lid of the mug and placed the mug behind the bowl of tea. The master taster would then be able to see the dry leaf, the infused leaf, and the brewed tea all at one glance.

We did this ten times a day, three days a week. We would taste every single tea, and as trainees we were supposed to taste more than just once. That is about three thousand cups of tea a week. You got used to it, or you left to become a merchant banker, as a couple of the trainees did. It was just too much for them.

Chapter 6

After being taught how to prepare tea, a trainee was then taught how to slurp tea from a spoon into his mouth, deep into the caverns of his tonsils, and then to swirl it around, breathing in at the same time. One had to avoid the cardinal sin of choking and spraying his tea master with a fine mist. It was an art, and it took weeks of practice and choking to finally develop my own slurp. Every taster has his own distinctive slurp, and after a few weeks I could tell which tea master was in the room without looking up.

The head tea buyer—an august, small man by the name of E. A. Locke, or "Lockie" to everybody except us lowly trainees—had a gentle sip, almost ladylike. Mr. Rossi, on the other hand, had a gargantuan slurp like a man truly enjoying soup with no one in earshot.

After preparing the counter, I would stand by the master taster, Mr. Mackie, as he would go down the bench. Four stars, two stars, burnt no good (NG)—I would mark the catalog and make note of the companies that would be interested in each of the various teas. There would be great excitement when an order would come in from Japan or America, as this meant that the best quality teas were to be bid on. He would tell me what teas were good and why they were good. He would smell the dry leaf and the wet leaf and taste the brewed tea—all in the space of thirty seconds. He'd make a comment, then move on to the next one. I was supposed to note the comment, write it down, taste the tea, then move to the next tea.

In my tasting room in Boulder, Colorado, I have the exact same setup—same bowls, same lids, same mugs, same scales. I have been to tasting rooms all over the world; they all have the same equipment. It is identical. We can compare notes with other professionals a continent away because we prepare the tea the same way. Tradition is a wonderful thing.

Lipton Training

A typical tea tasting.

Each tea had a code name. "Gull," "Quail," and "Eagle" all meant a certain grade and quality. Each Friday the cables would be sent out to the buying companies: 500 Gull available at 610, meaning five hundred chests of tea of Gull quality at a price of six shillings and ten pence.

On Monday, the orders would be waiting for us, and the trainees and the buyer would leave for the London auction to bid on the teas. The auction would be held at Plantation House on Mincing Lane, the home of the tea trade. The auction room was a two-storied amphitheater surrounded by the crests of foreign lands from whence came the tea. The auctioneer would sit in the pit, with a jobbing broker on either side of him to take bids from the smaller buyers. We would sit behind our buying brokers and lean forward to whisper our bids. This practice was supposed to keep secrecy intact. The trainee would keep a record of all the prices

Chapter 6

for each lot and the prices that the buyer paid for the lots we purchased.

I had to keep a running average of the prices (this was before the age of calculators). The lots were sold at about five a minute—it was fast. Panic would set in if you fell behind, because it was impossible to catch up. Mr. Mackie was a nervous buyer. He had been the managing director of Lipton Ceylon, but for some reason he had been recalled and was now a simple buyer. He would twitch and roll his eyes as he bid on teas and would constantly ask for the average. His broad Scots accent would thicken as the other buyers bid him up.

It became clear to me after a few weeks of attending the auction as an assistant that the auction was a big game to the large buyers, such as Lyons, Brooke Bond, and Co-op. They had the resources to bid up on teas and stick some poor buyer with a lot that he could not afford. Breaks of tea would come in lots of sixty chests or more. If four buyers wanted some of the tea but each buyer wanted twenty chests, there was not enough to go around, so the buying bidder would say, "Twenty Lyons, twenty Brookes, twenty Co-op; Lipton, you are out." This meant that Lipton had to bid up the price to see if one of the other buyers would drop out. If they all dropped out at the same time, Lipton got stuck with sixty chests of tea when they only wanted twenty. Everybody would snicker. It was all harmless, really. After the auction, private deals would be made, and the forty unwanted chests would easily find a home.

At the end of the auction, Mackie would sigh with relief. What seemed fun to me was to him an ordeal. The next day, we would receive the contracts. The trainees would have to go through them and check the contract against the catalog to make sure the price and the lot were correct. That afternoon, a half-pound sample of the tea we had purchased would come in (the purchase sample).

Lipton Training

We would taste it against the offer sample and then box the purchase sample. When they were all tasted and approved, we would make a bulk-up of the teas and send out a small sample of Gull or Quail to the buying company. At the same time, we were tasting next week's teas. On Friday, the cables would go out, and the whole process would start again.

London was the best place to train for a career in tea in the early '60s. The London auction was a powerful center. Teas from all over the world would be sent there for sale. In one place, you could taste Assams, Darjeelings, Vietnams, Indonesians, Ceylons, Japanese oolongs, South Indians, and even Turkish teas.

I lived in the northwest of London and took the underground every day. I became a commuter—one of the nameless many who swayed in the train as it made its way into the center of London. I was surrounded by gray and blue suits, morning newspapers, and umbrellas. I was part of the masses. It felt very strange. I would sit or stand and dream of Kenya or some other far-off land where they grew tea and hope.

My destination was Liverpool Street Station, a looming, glass-paned survivor of Hitler's bombing. I would walk up Liverpool Street to the tea warehouse on Bethnal Green Road in the East End of London, where all the Cockneys lived.

Sadly, the London auction is no more. It ceased to exist as the countries that grow tea took control of their own destiny. I experienced the end of hundreds of years of tradition and history. I felt honored to have been a part of the old trading ways. I still do.

For two years, I worked at the benches in London, switching between the North Indian counter, the blending counter, and the Ceylon counter. I was constantly learning more as I went along. E. A. Locke, he of the ladylike sip, was a giant in the industry. Locke was a man of few words but a wicked sense of humor.

Chapter 6

He once came out to taste a batch of teas I had prepared and started immediately. The only problem was that the spittoon was at the far end of the bench. He started to give his comments on the teas, which of course I had to write down, thus not giving me enough time to dash down to the end of the counter to get the spittoon. Locke happily went down the entire batch, spitting out the tea onto the wooden floor, a wide grin on his face. I never left the spittoon at the end of the counter again.

After two years, I felt I had gained all there was to learn at the counters, and I wanted to go overseas; but no work visas were available. Newly independent countries were looking to hire and train their own tea men. I had to wait. Lipton offered me a job in their supermarket division, but I chose to resign and seek my own fortune.

My mother came over from Kenya for a visit. It was the first time I had seen her in five years. I greeted her at the airport with a single rose. She had not written since I had last seen her, and she hadn't offered to help me get back to Kenya for my sister's wedding two years earlier. The wounds were deep but papered over. We toured Ireland together, and it was a time of reconnecting. My mother told me to come back to Kenya and offered to pay for the ticket.

I left England with little regret. We flew on a charter airline that declared bankruptcy as we were in the air. We landed in Nairobi, and my father hugged me. I was home. I was back in Africa.

Chapter 7
The Brothel

When I arrived back in Kenya, my sister and her husband had just moved into a new house. They mentioned to their landlord that I was returning to Kenya; the landlord and his company in Mombasa were looking for a tea taster.

A few days later, I was flying in a small, single-engine plane down to Mombasa. The pilot, Peter Winch, was the owner of the tea and coffee company. Winch was a larger-than-life figure—rotund, bearded, and loud. A Kenya character.

"Lean forward as we take off," he yelled as we sped down the runway at Wilson Airdrome. "I think we are a touch heavy." He said this as the end of the runway rapidly approached. We cleared the fence surrounding the airdrome by a few feet. "Ahh . . . up at last," yelled Peter, clearly enjoying my discomfort.

We landed for lunch at Mtito Andei, the halfway point between Mombasa and Nairobi. Peter angled the plane down and swooped over a game lodge. "Who's hungry?" he yelled.

A couple appeared on the steps of the lodge, and Peter waggled the wings and then swept low over the dirt landing strip. "Just chasing off the warthogs. Last trip, damn near hit one. Would've ruined lunch."

Chapter 7

We landed, and a Land Rover appeared with the couple driving to haul us over to the lodge. I was back in Africa with a vengeance.

Mombasa was hot and steamy, and the approach over the putrid swamps was bumpy and eventful, with Peter yelling at me to open the door as we rolled down the runway. "Got to slow us down, old boy!"

The next day, I was picked up and taken to the offices in the industrial area to meet the tea team. The shortest job interview in the world is for a tea taster.

"Here are twelve teas. What are their origins, what are the grades, how much are they worth on today's market, and who would use them?"

You either know what you are doing, or you don't. I did. I got the job.

After the gloom of London, Mombasa was a joy. I moved into The Castle Hotel, which offered me a three-month stay at reduced rates. Breakfast, lunch, and dinner were part of the deal. The Castle was a large white hotel on Kenyatta Avenue. Scalloped windows sat beneath a turreted roof. It had pretensions of grandeur. It was also the local brothel of choice.

The girls hung out in the arched patio bar, which stretched from the foyer to the street. Sailors, Arabs, locals, visiting businessmen, and all manner of brothel clientele would pay courtesy stops. The girls made their deals and took their customers upstairs.

My room was on the second floor, facing the courtyard. After a while, I became a fixture, and the girls lost interest in me as a potential client. My mother came to visit and stayed at the hotel. She was mistaken for one of the girls, and an Asian gentleman took a lot of convincing that she was not for rent. After my mother's visit, I became part of the fraternity, included in their chatter and gossip.

The Brothel

> ## Tea Tales
>
> Mombasa was, and still is, a major auction center. Teas from Kenya, Uganda, Rwanda, Burundi, the Congo, and Tanzania all made their way down to Mombasa to be auctioned.
>
> The main tea-growing region in Kenya in the late 1950s and early '60s was in Kericho, a good day's drive from the capital, Nairobi. Most of the estates were owned by large corporations located in Great Britain, like Brooke Bond and Finlays. Around the early 1960s, the British government started the Kenya Tea Development Authority (KTDA), which was initially a smallholder project. This meant that small farmers would grow approximately an acre of tea, and they would harvest every seven days and take the tea to a central processing plant where they would sell it. This scheme became one of the most successful tea projects ever and has propelled Kenya to become one of the largest tea exporters in the world. In 2002, they produced over two hundred and fifty thousand metric tons of tea and exported most of it. By expanding the tea fields in the '60s, the KTDA was able to make use of the most modern technology available at the time.

I learnt about the foibles of mankind: the elderly gentleman who took so long he was prepared to pay double; the young man who was so quick he always tried to negotiate a discount; the seamen who wanted to double up, and sometimes triple up, on a girl; the local business leaders who would slip in the back and call down to the front for their girlfriends. I met a few of them on the staircase.

Nothing is stranger than two men meeting in the stairwell of a brothel. They both presumably know why they are there. This provides instant simpatico and comradeship. But the nature of the business requires a certain formality. There were those who slunk

Chapter 7

around, face averted, and those who strode up the center of the staircase. Often it depended on whether they were walking up the stairs or coming down. I gained a certain reputation until someone discovered I lived there. It caused amusement.

I learnt that the girls were all concerned with the mundane issues of life—enough money for food, for clothes, for their children, their boyfriends, sometimes their parents. They had cliques and friendships and were united only in their hatred of the police and customers who tried to cheat them. They had their favorites and their regulars. To sit with them and listen to their chatter was to be included in a rare group. It was like being surrounded by a flock of exotic birds.

They were young girls from a variety of tribes, yet they created their own tribe. Their tribe had its own customs. Woe betide a member who tried to steal a regular customer from another. Or offered herself at lower rates. Or stole. They had their own code, their own sense of honor. When the fleet was in town, they would band together against the hundreds of young women who would come down from up-country to service the sailors. There would be catfights and rivalry between the Mombasa regulars and the interlopers.

The Castle Hotel is still there, a sad sight now: boarded up, the brilliant white exterior gray and stained with the dripping of broken gutters. When I was last there, I peered through boards at the cracked patio. It was covered in debris and weeds. I remembered the gaiety and the risky sense of fun. It was better that way. AIDS has devastated Africa now; careless sex is no more. It would not be possible to even think of living the way I did then. I hope all the girls survived.

I was sad to leave after my three months were up, but I had been told that it was inappropriate to live in a brothel. I moved in

The Brothel

with a young bank trainee in his bank-provided house. The house was in Nyali, an upmarket residential area and a very appropriate address. But I missed the girls and would go back and have a beer with them. They always welcomed me with high-pitched giggles. I was never treated as a potential customer but as a returning member of the tribe. I felt honored.

UGANDA

Chapter 8
Uganda Bound

In the '50s and '60s, Kenya was still young and full of eccentric characters who made a point of their eccentricity.

Known as the dumping ground of the English gentry and their second and third sons—and occasionally a ne'er-do-well first son—in the '20s and '30s it was still a land of adventurers, rascals, scoundrels, businessmen, soldiers, and white hunters. The white population of Kenya never amounted to more than one hundred thousand people, even at the height of the settler movement. Our time was doomed even as we lived it.

I entered a culture clash in the small Kenyan tasting room. The three of us who worked there—Don Lowe, Mick Clark, and I—were like oil and water. Don, who was precise, neat, and meticulous in every detail, was destined to perish in a terrible accident many years later when he was thrown from a car. Mick Clark was a quintessential trader, wheeling and dealing. Finally there was me, for whom details have always been an issue and playing rugby was more important, quite frankly, than staying late at work. For six months we tried to make it work. Then Peter came to me and asked me if I would like to move to Uganda to open up the country and establish a trading office. I was twenty-three years old. I didn't have to bother thinking before I said yes.

Chapter 8

I left a week later. I had four hundred sample bags, six tasting cups and lids, a tea scale, my tasting spoon, and an optimistic outlook. I stopped in Nairobi to see my family; it was as dysfunctional as ever.

I stayed for a week in Nairobi then moved on. It was a three-day drive from Mombasa to Kampala, the capital of Uganda. The border was at Jinja, where Lake Victoria empties into the Nile River. The drive through the northern part of Kenya was dry and dusty. Endless vistas of gray scrub stretched before me under an arched blue sky. Cross the border, and you were in a new world of greenery: lush vegetation, banana trees, soaring jungles, and miles of water-saturated papaya grass. The road runs like an arrow, straight through the heart of the papaya swamps. A narrow ridge of a dike ends in a shimmering mirage during the day and pitch-black darkness at night.

The sounds are different in Uganda. There are more birds, the air is wet and slightly musty, and the heat of Kenya gives way to a languid softness. It is a more violent country; gangs roam the land. A lone European is easy bait.

I arrived in Kampala at dusk. There was a brilliant sunset and then the sudden darkness of Africa. The city sits astride seven hills, and the roads curve in and around and up and down. I drove through the clatter that is the evening and made my way to the Apollo Hotel, high above the town.

George Kargarottas, a barrel-chested Greek, was there to meet me. He ran the coffee-trading business for Peter Winch, and I was to run the tea. George was the one issuing my paycheck, so I took him to be the boss. He responded well to that approach.

"Have a beer, and tell me what you are supposed to be doing up here. I know nothing about tea, except it's grown in all the dangerous places around here."

"Dangerous, huh?" I said, trying to sound very nonchalant.

"Very!" said George, taking a deep draft of his beer and looking over the rim of his glass at me. "I think you should hit the road tomorrow and go up to the growing region, stay there for a couple of weeks."

"Tomorrow, huh?" I replied, trying the nonchalant approach again.

"You should learn the lay of the land and the growers, drive around, get yourself comfortable, learn the ropes—where to go, where not to go."

"Where not to go, huh?"

"You were raised in Kenya, right? Speak the language, right? Know Africa, right? Can handle yourself, right?"

"Ah . . . sure," was my only response as I buried my face inside my glass and sought some Dutch courage.

After about six large beers, I vaguely remember telling George that I was ready to leave that night—a get-started-early type of thing—and George telling me that tomorrow would be fine. He also told me the hotel was full of loose women. He had heard about my stay in The Castle, and he let me know it would not do to have that reputation in Uganda. I remember trying to explain the circumstances, but he left before I finished.

The next day found me on the road, bleary-eyed and feeling distinctly queasy, heading for the dangerous tea-growing areas.

Chapter 9
Working the Fields

After a harmless, uneventful drive along the shores of Lake Victoria and then on into the foothills of the Mountains of the Moon, I drove to the Munobwa tea estate. It was owned by Derek Broadhead-Williams, an old friend of Peter Winch's, and I was expected.

I was taken to my bungalow, which was on the top of a steep slope. Densely packed tea bushes brushed up against the side of the porch. I was told that my day started at five a.m. at labor lines.

"Labor lines" was the morning roll call for the labor force that picked and made the tea. Rather like an army muster roll, the names of the laborers were called out, and they answered "*aya*" or "*dio*" and were marked present. Then the division was made between the field crew, who picked the tea, and the tea makers, who worked in the factory.

My first week was spent working the fields, seeing how tea was picked, collected, and weighed. Making tea is a business. Tea estates are supposed to make money, and the green leaf is the beginning of the process. Green leaf is laden with moisture; it is very heavy, but it must be weighed. The pickers are paid based on how much they pick, so they were very conscious of the weight. Some would

Working the Fields

happily add stones or the odd rock or two, and it was my job to make sure that what was weighed was all green leaf.

A normal day's work was about 60 kilos, or roughly 120 pounds of leaf. This is a lot. I tried my hand at picking, and it was hard work. The wicker basket was heavy and hurt my head; the two leaves and a bud, far from being easy to pick, were so supple they were difficult to snap. The bushes were so dense that it was tough to force your way through them. After I'd collected half a basket, I handed the job back to the field supervisor, who had been watching me with an amused grin on his face. I headed back to the shade, suitably humbled.

In the field I learned how to get upwind of fertilizer applications and to wear boots in the long grass. This was after I felt a funny feeling on my legs one morning, a kind of itchy, slimy feeling. Looking down, I saw little curly animals—funny little things—crawling up my legs, and as they got higher, they got fatter. Leeches! Prancing around, I tore off my trousers. Naked from the waist down and surrounded by a laughing crew of pickers, I plucked off the leeches, ripping my skin on each one but not caring.

All the pluckers were male; that is one of the oddities of the estate world. In Asia, all the pluckers are women; in Africa, they are all men.

Boots were pretty useful to protect against the snakes that occasionally would lie and bask on the dry, exposed roots, coiled and relaxed until you stepped on them. The comfortable feeling of boot leather up around your ankles was welcome.

The teas grown in Uganda were medium-grown. They were not prized for their extraordinary flavor but more for their color and strength. They were manufactured to accentuate these characteristics.

Chapter 9

Tea Tales

The ancient or "orthodox" way of making tea is by far the slowest and most gentle method. It is so gentle, in fact, that the fine hairs that are on some of the tea bushes are not removed during the processing. When the tea is dried, the hairs turn a golden color. This type of tea leaf is called "flowery" because of its style and "tippy" because of the fine hairs. If the color of the hair after drying is gold, it is called "golden."

There is another method of making black tea. It is violent, fast, and very effective, producing a lot of similar-tasting tea very quickly. This method is called "cut, tear, and crush" (CTC). I also have heard it called "cut, tear, and curl," but to be honest, I do not see where any curling could take place.

The beauty of the orthodox method is that it produces lots of different sizes of tea and lots of different tastes. Orthodox teas can be graduated into many grades and sub-grades. Not so with CTC; this method is efficient. With CTC, about 80 percent of the leaf ends up as one grade: pekoe fannings. It is even and small, and it brews up with great color and intensity. The other 20 percent of the CTC manufacture is split between broken pekoe and dust.

The ancient way of making tea is to roll the tea leaves after they have withered and the natural moisture has been reduced to about 40 percent of what it was on the bush. At that point, the tea is pliable and can be rolled to a tight twist, which is the classic pekoe black tea.

If the pekoe broke during rolling, it was sifted out and became broken pekoe tea. If it broke yet again during rolling, it became quite small and fell to the floor. At that point, it was collected by the tea makers by waving fans at the tea and collecting it all in a corner; this tea was called fannings tea, or broken pekoe fannings. Anything smaller than fannings was called a dust, for obvious reasons, but was still collected and sold.

In simple terms, these are the primary grades of tea—pekoe, broken pekoe, fannings, and dust. You can see that these grades all relate to size and not taste, although the smaller the leaf, the stronger the taste. Dust, therefore, is always much stronger than broken pekoe from the same estate.

Tea professionals who read this will wonder what happened to the word "orange." Should not the grades read "Orange Pekoe," "Broken Orange Pekoe," and "Broken Orange Pekoe Fannings?" The answer is "Yes!" But I chose not to include "orange" in these descriptions. For the curious, the legend goes that the Dutch named their teas in honor of the royal family from the House of Orange. Over the years it stuck and now most orthodox primary grades have this meaningless word incorporated into their titles.

When it came my time to serve in the Munobwa factory, I was entranced by the rich, vegetative smell that permeated the building; it was fresh and juicy and smells healthy.

The leaf came in from the field; it was carefully weighed and then weighed again because it had already started to lose moisture. It was placed in long troughs with chicken wire at a depth of three feet; cool or warm air was then blown through the troughs and up through the leaves to control the pace of the "withering."

In the old days in Ceylon and India, they used to place the freshly picked tea leaves on multiple racks in the lofts above the tea machinery. These lofts were surrounded by hundreds of windows, which were opened to breezes that would naturally wither the leaves. This system took up to seventeen hours to complete. This accounts for tea factories in India and Ceylon being built on top of ridges, to catch all the breeze that was available.

Chapter 9

A factory in Ceylon (Sri Lanka) with "withering lofts" up top for drying tea.

It also meant that tea factories were very imposing buildings, masses of shuttered windows, painted white; they lorded it over the surrounding valleys. You get the sense, as you drive up the winding roads to the factory, that you were meant to feel small and insignificant—a lot like the cathedrals built in the Middle Ages or perhaps buildings in Communist China might make you feel. These were huge structures designed to make mortal man feel mortal.

At Munobwa, once the leaf had been withered it was taken by conveyor belt directly to the rotorvane machine. The rotorvane was essentially a huge sausage maker; you fed the leaf in at one end and a spiral feed took it up through a tube, crushing it as it went, so it popped out at the end in a semi-macerated state.

The leaf was then fed directly into the cut, tear, and crush (CTC) machine. This machine consisted of two rollers with razor-sharp ridges running vertically so that they interlocked. The leaf pieces, already chopped up by the rotorvane, would fall into the CTC rollers and come out the other side in a stream of shredded tea.

Most factories had a bank of three to four CTC machines. Munobwa had three, and after going through all three, the leaf had been reduced in size to something resembling a pinhead. All of this processing would take ninety seconds from the first CTC cut to the last. The macerated leaf was then fermented in trays and dried in a wood-fired dryer.

This last item was a monster, and it had two issues. The first was that it required a lot of wood to keep it fired up, and the second was that if it leaked smoke into the drying chambers, the tea took on a smoky taste that ruined its value. Part of my task in the factory was to check the dryer for leaks.

Seeing as the temperature was always set at about 270 degrees Fahrenheit for the feeder mouth, this was hot, unpleasant, dusty work. But that's what trainees are kept on the estate to do, and I learnt to love the rhythm of the days, the cool nights, and the blinding-hot dryer room. When it came time to leave the estate and get on the road to develop the trading business, I was filled with a sense of regret. The plantation life was one that appealed to me—I enjoyed the routine and the fresh air—but my future lay in a different direction.

Chapter 10
Ahead of His Time

I**N A LAND OF ECCENTRICS, FRAZER SIMPSON WAS A LEADING LIGHT. HE AND HIS BROTHER RAN THE KIJURA TEA ESTATE. AS AN INDEPENDENT ESTATE, IT CLEARLY WAS A SALES OPPORTUNITY FOR ME. HE QUICKLY BECAME A FRIEND AND THE MOST ENTERTAINING OF COMPANIONS.**

The Kijura estate was located in the foothills of the Mountains of the Moon. You reached it by a torturous road that was full of potholes and stones and surrounded by tall grass so that you could not see around any corners. On my first visit to introduce myself, I stayed in town, but I was soon invited to stay at the estate the next time I was in the area.

The tea trade in those days included well-established companies such as Mitchell Cotts, which had large holdings that no trader could get near. Other estates were owned by Indian groups that were very suspicious of traders in any form. Finally, there were the independent estates hacked out of the virgin bush by intrepid settlers; these settlers raised their families and had hopes of a better life.

Kijura was one such independent estate. Frazer was the younger brother, and he used to drive his elder brother John to distraction. Frazer liked to tinker and was constantly changing the parameters of the production in a never-ending attempt to improve the quality of the product. The trouble was that he changed not just one element at a time but many, so when he did produce a good batch of tea, he did not know which change was the one that caused the improvement.

Dinner with the two of them was hysterical.

"Changed the withering temperature, I saw." That came from John.

"Yes, and added another six inches of leaf." That from Frazer.

"Got a better wither?" From John.

"Yes, I think so." Frazer.

"What caused it?" John.

"Could have been the extra six inches, or the temperature change." Frazer.

My most memorable wakening was at seven a.m. after a night of hospitality from the brothers. I awoke to Frazer pounding on my bedroom door.

"Barry, you must come. I've done it!"

I joined him moments later, boots on, shorts on, shirt going on, no need for a hat. When Frazer said he had "done it," best to go see what he had done.

His racket had awakened John, who emerged tousled and equally bleary-eyed.

"What's up?"

"I've done it. I've made a dramatic improvement." Frazer, when excited, tended to hop from foot to foot, going at it like a demented dervish.

"Stop hopping, Frazer, and tell me what you have done." John.

"Show you," and with that, Frazer stopped hopping and shot out the door. We heard his jeep start up and roar off up the factory road.

Chapter 10

"Better follow him; no idea what my bloody brother has done." John left to get his car keys, and moments later we joined in the drive.

In the factory, sitting in the center of the fermenting room, was a refrigerated Coke machine with a power cord coming down from an electric light socket. The front of the Coke machine had been cut open, and a reed basket had been jammed in and stuffed with fermenting tea. A low hum came from the Coke machine.

"Chilled fermentation. That's the answer." Frazer resumed the dervish dance. "Come, Barry, taste it." He shot off into the tasting room, where three teas had been prepared.

I tasted them. They were superb. Wonderful color, fabulous flavor, better than anything I had ever tasted from a Uganda tea estate.

I gave my professional opinion. "Wow," I said.

Frazer's hops elevated another inch or so. "Wow," he said. "Wow, I knew it. I knew it. It's the answer."

John loomed in the doorway. "Frazer, how much tea does that Coke machine hold?"

"I don't know—couple of pounds."

"How much do we produce in a day?"

"About five thousand pounds, but I've thought this through. We need a large commercial refrigerator unit that we can run like a dryer, keep the teas in there for two hours, and then dry them."

John raised his eyes to the ceiling. "Frazer, do you know how much they cost to buy and to run? The electricity cost alone would break us."

Frazer stopped hopping. I felt he needed a boost.

"These are damned good teas, Frazer, really good." But I could see my word had not worked any magic.

Ahead of His Time

Fifteen years later, I was touring a tea factory in Kenya. A large, elevated conveyor belt carried the CTC leaf and dropped it into a machine that had blowers running down its side, blasting cold air over the fermenting leaf, which was slowly moved from top to bottom.

"What's that?" I asked, knowing the answer before it came.

"That's the latest in chilled fermentation; the leaf has cold air blown on it. Produces very good liquors."

Somewhere in the universe I thought I heard the distinctive sound of someone hopping.

Chapter 11
Guns and Golf

LIFE IN UGANDA BECAME INCREASINGLY DANGEROUS. THE POLITICAL SITUATION WAS UNSTABLE, AND IT BECAME SENSIBLE TO BE HOME EARLY RATHER THAN LATER. BUSINESS BEGAN TO SUFFER AS ONE OF MY PRIME ADVANTAGES BEGAN TO SLIP AWAY—TIME.

I had developed a system whereby I received samples and offers from estates like Kijura and Munobwa and offered contracts for them overseas via telex on a "subject to approval of sample" basis. This meant that I had tasted the teas, they were okay, and if the buyer did not like the sample I sent him, he could cancel the contract. All of this would take place as the teas were being packed on the estate or as they were on their way to Kampala.

If I sold the tea before it got to Kampala, I was able to divert it to our warehouse, put it on a railway wagon, and send it directly to the ships in Mombasa. This procedure was saving a lot of money for everybody and giving the estate a much higher return than they would receive from going to auction in Mombasa with all its fees, or even worse, being sent all the way to the UK to go to auction there.

But with the advent of the security situation, samples began to take longer and longer to get to me, and even my trips to the estates became more and more risky. It was clear that the country's situation was coming to a boil. One evening I was stopped while returning to Kampala after my twice-monthly visit to the tea estates nestled up against the Mountains of the Moon.

I had been in Uganda for two years. Twice a month I would take the six-hour drive up to Fort Portal, collect the tea samples, drive around and visit the estates, have a game or two of golf and dinner at the club, and then head back to Kampala. My treasure trove of fresh tea samples sat on my rear seat, well away from the gasoline fumes in the trunk.

It had been a normal day when I left Fort Portal. I was slightly hungover from an evening with two local planters. The drive was uneventful until I rounded the corner and was pulled up by a crude roadblock. Chairs, a sofa, and an upended table stopped me. I was the only car at the block; in fact, I was the only car on the road that I could see.

Six soldiers manned the roadblock. All of them had their eyes on me, their guns loosely pointed in my direction. What I did not know was that someone had tried to assassinate Milton Obote, the president of Uganda, and a curfew had been imposed. I had been on the road and had not heard the news.

The short, stubby snout of the machine gun was about six inches from my face and looked like a railway tunnel.

"What are you doing here?" The sergeant's voice was harsh and demanding.

"I am driving to Kampala," I replied, keeping both hands on the steering wheel to stop them from shaking.

"Where are you coming from? What are you doing here?"

"I came from Fort Portal. I am a tea taster." I looked up into the eyes of the sergeant holding the gun, trying to connect, trying to make myself human to him.

Chapter 11

> ### Local Lore
>
> Of all of the oppressive regimes in African history—and there have been a lot of them—Idi Amin's is considered to be one of the worst. During his eight-year reign, he is rumored to have tortured and killed close to five hundred thousand of his fellow Ugandans.
>
> He was notoriously paranoid and prone to "visions," and some speculate that his erratic behavior may have been due to late-stage syphilis. It's certainly true that he acted strangely. At one point he declared himself King of Scotland, and at his death his most popular title was "His Excellency President for Life, Field Marshal Al Hadji Doctor Idi Amin, VC, DSO, MC, Lord of All the Beasts of the Earth and Fishes of the Sea, and Conqueror of the British Empire in Africa in General and Uganda in Particular."
>
> Suddenly "international tea master" doesn't sound quite so impressive.

"A what?" The gun jerked toward my ear.

"A tea taster." It seemed an absurd thing to say while staring at the mouth of a machine gun.

"What is in those bags in the back seat?" The gun gestured over my shoulder.

"They are tea samples."

"Tea samples? What are you doing with tea samples?"

"I was up in Fort Portal collecting tea samples. I buy and sell tea." I stopped talking and sat still.

The gun moved back a fraction. Uganda in 1969 was no place to be. Two years away from the coup run by Idi Amin, it was a seething ferment of suppression and violence. The gun by my face did not waver. "What do you have in the back of the car?"

"My suitcase and my golf clubs."

The gun snapped back into my face. "Clubs." The voice rose. "What clubs?"

Guns and Golf

I was stumped. How do you explain golf to someone who does not know the game? I looked up into the sergeant's eyes again. What I saw scared me. I was raised in Africa, and I knew the signs of anger, remorse, and, worst of all, panic. What I saw was panic.

"Can I get out of the car and show you?"

The sergeant stepped back, and I opened the door; the rifles suddenly swiveled in tight hands and pointed directly at me. I shut off the car engine, took the keys, ever so carefully walked around to the trunk and opened it, and then stepped back. The sergeant advanced cautiously and peered into the back.

My suitcase was resting on top of my golf bag. The sergeant gestured for me to remove the suitcase. I pulled it out and lay it on the ground. The sergeant looked at the golf bag and gestured at it.

I reached in and pulled a golf tee from the little fastener on the outside of the bag and held it up for him to see. He reached out and touched it, then replaced his hand on the gun barrel. I took that as a good sign and pulled out a brand-new Dunlop golf ball and peeled away the wrapper. He reached forward and fondled the ball for a moment, then stepped back.

I looked in his eyes. The panic had gone, and curiosity was creeping in. I held out the tee and ball and walked over to the grassy edge of the road. I stuck the tee in the ground, placed the ball on top, and stepped back, terribly conscious of the rifles concentrated on my back.

The sergeant looked at me. I walked back to the trunk and ever so slowly pulled out my driver. I looked at the sergeant and at the leveled gun, which had now moved up to my chest, and nodded at the golf ball.

He stepped back, and I walked over and faced into the jungle. I addressed the ball and took a swing at it, knocking it way off

Chapter 11

into the trees. I turned and watched a broad smile break across the sergeant's face.

The sergeant reached up and removed his gun sling, handed the gun to the nearest soldier, and reached for the driver.

"Ball," he said, nodding at the golf bag.

I watched him smack the ball off into the jungle and then roar with laughter. That was that. Every soldier had a swing with the driver until I had no golf balls left. The soldiers waved me on my way with the advice to get off the road as soon as possible because of the curfew.

As I drove away, I started to breathe again and wonder at the reality that was Africa. I could easily have died that day; instead, I made some friends, and I am sure they told the story of the mad Englishman and his golf clubs. But what if my life had ended there? And what on earth had led me, a tea taster, to be in harm's way? I decided that it was time to move on. Africa had my heart, but the next stop on my journey would be elsewhere.

Chapter 12
Waiting on a Visa

As if the universe had heard my decision, a week later Peter Winch flew up to tell me that he had little confidence in the future of Uganda and was closing down the operation.

He felt that I had done a great job and was giving me three months' home leave. He was sad to see me go but said Uganda gave him the willies. Three days later I was on my way to England with an overwhelming feeling of sadness for Africa, a beautiful land and a way of life that I somehow knew wasn't going to last. London beckoned, but then what?

The jungle telegraph of the tea trade had been at work. The industry was small. From grower to broker to buyer and blender, it was an intimate fraternity. Gossip about abilities and achievements and failures was the lifeblood of the trade. From club to boardroom to auction center, the trade thrived on information. If you were any good at what you did, it became known. If you were a hopeless failure, that too reached the ears of those who ruled the very small world of tea. The news that I was back in town spread quickly, and Lipton, my old company, heard. Much to my surprise I received a call from Lockie and an invitation to lunch.

Chapter 12

The meal was a quiet affair, held in the executive dining room. As a trainee I had dined in the employee cafeteria. It sat alongside the executive dining room, and every now and then we had tempting glances of white tablecloths and polished silver as we ate on wooden trestles off tin plates.

On this occasion I walked past the cafeteria, noting the trestles were still there, and entered the executive dining room with its tall arbored windows, hushed atmosphere, and soft carpet. Against the rear wall was a cabinet with tonics and a bottle of gin set next to a bucket of ice. Opened bottles of Bordeaux waited to be poured. With a gin and tonic and a couple of glasses of wine inside me, I had little trouble regaling Lockie with my tales of African adventure. Afterwards he invited me back to the tasting room.

"If you have been in Africa, you must have tasted a lot of African tea, then?" Lockie was his usual taciturn self.

Fueled by the lunch, I nodded. "Lots and lots."

"Thought so," said Lockie. "Well," he said, tipping his head at the bench, "in there are a bunch of Ceylons and Indians; go taste 'em and tell me what you think."

I sobered up immediately. This was no joke. I knew the Lockie of old; he had provided lunch and tolerated my stories for the sole purpose of this moment.

I slurped and spat my way down the teas, commenting on each one, relying on instinct and first impressions—always the best way to go with tea. At the end of the batch, I reviewed my tasting comments and thought I had got most identified and valued correctly. There were a couple of teas that struck me as cuckoos—teas that were not supposed to be in a batch with those origins but were included to trip up the taster. I thought one of them that had a very distinctive flavor and character was Kenyan orthodox tea from the Michimukuru estate. The other was an Indonesian that tasted a lot like a South Indian. I had noted these two in my remarks. I would be either very right or very wrong.

I waited as the assistant who had jotted down my notes took them into Lockie's office. After a few moments, the great man emerged, head down, hands in his pockets.

"Found the birdies then, did you?" He looked up at me. "We did a good job training you. I think you'll do for our American cousins."

> ### TEA TALES
>
> A professional tea taster does more than just taste the tea. He or she must also note important aspects of each tea, such as leaf style, brightness, color, astringency, and origin. Like wine tasters, tea tasters have their own rich vocabulary. A fruity, brassy, chesty tea is bad; a full-bodied tea with pleasant favor and bloom is good.

I looked over at Lockie and saw the flicker of a smile.

"We might have a job for you. Mind going to the States?" he inquired.

"Wouldn't mind a bit," said I.

Three weeks later, I received a letter inviting me to the office to meet a Mr. N. F. H. Fleming, or "Toby" to his friends.

Many Englishmen have multiple names; it has to do with family and tradition, but Toby's name is something else. To go through life answering to Noble Fernley Hutchinson Fleming, you have to have style, which Toby certainly did. He was a seagoing man, ex-Indian Navy. He was also ex-cavalry and a member of the Calcutta Light Horse. He had a gruff voice and a hooked nose set off by twinkling eyes and mannerisms that greatly endeared him to all who worked for him. Toby was director of tea buying for Thomas J. Lipton, an Englishman in charge of the largest American tea

Chapter 12

company's buying department. Our first meeting went well, although I heard later that he had suggested I get a haircut. He asked if I was interested in joining Thomas J. Lipton. I accepted the offer and prepared for the journey, while Lipton applied for my American work visa.

All was not sunshine and light. The Immigration Department decided that there was no such profession as a tea taster and rejected my visa application. This would have come as quite a shock to the tea tasters who were laboring away daily in the United States—men at Tetley USA, Brooke Bond USA, and Lipton USA—not to mention the Food and Drug Administration's official tea taster, Mr. Robert Dick. They would have been surprised to hear that their profession had ceased to exist. But the immigration department is a law unto itself. They ruled that there was no such profession and considered the matter closed. Nonsense, said Lipton, who set out to prove there was such a calling. I was told the visa was going to take a little time, and in the meantime, they had fixed up some work for me with a few London tea firms to expand my horizons.

The year was 1970, and the London tea trade still had the trappings of its former glory. The selling and buying brokers had moved from Mincing Lane, the home of the trade for hundreds of years. The trade had moved en masse to Sir John Lyon House at High Timber Street, alongside the Thames River down toward Tower Bridge.

I was to join the very respectable selling broker Wilson Smithett & Company, a firm with extensive ties to Africa, Malawi in particular. I had worked with many of the brokers and knew their procedures.

I was given the job of marketing estate teas. This involved striking up direct correspondence with the estates so that I could relay information to the buyers about weather conditions and anything that might help sell the tea. More importantly, I was invited to join in the social activities when the estate owners came to London to meet the brokers, buy new machinery, and find wives. The search for a wife was particularly intense.

The life of a young planter was very active, but it could be lonely on the estate. There was always the temptation of the local ladies, but that could be flirting with disaster. Of course, wives of estate managers could be courted in a chaste, knightly fashion, and when their daughters visited from boarding school, they enjoyed intense competition and attention. If they were of marriageable age, the chase was hot and heavy. These were girls who knew the tea business and the hardships and freedoms of life in the hills. Perfect mates for young planters.

But these girls were few and far between, and there was simply not enough suitable bride material in the remote tea areas. When a young planter had sufficient funds and time, his thoughts turned to marital bliss. Talk in the Hill Clubs revolved around suitable mates, and the search for one became rather like a war campaign: decide upon a strategy, reconnoiter the terrain, prepare the ambush, capture the prey, and carry her off to a distant tea plantation to live happily ever after. Sometimes it worked.

Most of the planters I knew were either English or Scottish, so the preferred hunting grounds were London or Edinburgh. The pursuit was glamorous and involved a lot of wining and dining. The objective was to sweep the girl up in the rush of events. A life of tennis parties, club socials, two cooks and a gardener, not to mention a housekeeper, maid, and driver, must have sounded like heaven to these young women. Many a hushed "Yes, I will" resulted.

The life they headed off to was not a bad one. It changed over the years. It later became more dangerous as tea-growing countries

Chapter 12

suffered the pains of independence and the lack of security that often came with it. But even today when I visit the estates, the life of a planter is one of early mornings and hard work but a relaxed, well-catered home life.

The young planters did not have much time and were very focused. One day I was sitting in the office of Wilson Smithett & Company when an assistant estate manager arrived on his first day of a three-month home leave. After the formal introductions, we began to talk.

"Know any Scottish nurses?" the young planter asked.

"Nurses . . . uh, not really. Why? Is something wrong?" I replied.

"No, everything's fine. Looking for a wife."

"Oh, I see. But why a Scottish nurse?"

"Nurses make good wives, something to do with being around all those sick people, I suppose. Scots can put up with hardship. Anyway, looking for one."

"Sorry, can't help you, don't know any."

"Not to worry, I'll go down to Guy's Hospital and do a little recon. Bound to find something wandering around."

I heard later that he had.

Sir John Lyon House is no longer the center of the tea trade. The introduction of the overseas auction centers in the 1970s finished off London as a tea center. The London tea trade was once the pride of the tea industry; it was the place to learn about tea. Such is no longer the case. Even Wilson Smithett was put up for sale in 2005. Now the major packers buy the tea directly from the estates or from the auctions at origin. With its demise passed an icon of my youth.

I was sad to leave the brokerage business. I had became adept at enjoying lunch, but the day came when Lipton told me my time at Wilson Smithett was up. My next assignment was to join a company called Brash Brothers. It was a small, private-label company that packed teas for other companies. The Brothers had many blends, and my task was to match the blends from the stock of tea on hand. It was a challenge I enjoyed.

The issues in matching a blend were quality and price. Imagine you are a baker and sell cake to a customer who then requests that you deliver the exact same cake every week for a year.

"Great," you say. "Steady business." Then the price of flour doubles. Then you discover that you cannot obtain the same type of sugar, and the milk you use is no longer available. You still have to bake the cake, but you have to use different sources for the ingredients, which are likely to taste different. But you have to make the cake taste the same … every time! Not only that, but you have sold it at a set price, so you have to make sure that it costs the same every time as well.

That is the challenge of tea blending. It is a constantly shifting scenario of availability and price. Tea quality changes with the seasons, and prices change weekly as the auctions around the world define the prices. Droughts, strikes, and devastating monsoons can all create havoc with supply and pricing. But you still have to get your supplies in to sell that damned cake.

Some origins are close in color and taste but not close enough to be able to interchange easily. Kenyan teas have a red color and a strong taste, as do teas from the Assam Valley in India. But the taste differences are sufficiently wide to limit the use of either as a total substitute for the other.

Indonesian teas, with their light golden color, can be used to help out a shortage of Ceylon teas in your stock, but they lack the high-grown astringency that Ceylons possess. So you have to add a pinch of something else astringent. Let's say you use the most

Chapter 12

> ### Tea Tales
>
> A tea taster will use a large spoon and noisily slurp the tea. This allows more oxygen to pass over the taste buds and gives a better sense of the tea's flavor profile. Once the tea has contacted all parts of the taster's mouth to ensure an accurate evaluation, it is expectorated into a spittoon. A professional tea taster can be expected to taste hundreds of teas a day, and using the same spoon ensures that all the teas are treated equally, since the material and shape of the spoon can affect taste.

astringent tea I know—a Darjeeling. You have the taste balanced, but you just blasted your costs out the window because Darjeeling teas are very expensive.

I loved the blending game, the challenge to match each blend and keep it within cost. It was fun, and once something becomes fun, it is no longer work. My three months at Brash Brothers passed quickly.

Eventually the call came from America that my visa had come through. As a farewell gift, Brash Brothers presented me with a tasting spoon. I have it to this day—aged, battered, and worn, but still serviceable. Together we have participated in a million cups of tea.

Chapter 13
American Ball Busters

Lipton had proved to an immigration judge that the tea-tasting profession did exist. I found out later that, while the immigration judge had finally agreed there was such a job as tea tasting, he was not prepared to add it to the list of approved professions in the Federal Register. The profession was simply too small to matter.

As it turned out, I was allowed into the country under a special type of visa, the same one that defecting spies and other persons of "high and unusual value" to the United States were given. I decided it best not to ponder which part of that crowd I best fit into.

My first night in America was spent on Route 4 in Fort Lee, New Jersey. It was gasoline alley: nothing but cheap motels, gas stations, and an eight-lane highway that roared twenty-four hours a day. In the midst of a deep sleep, I was awoken by a knocking on my door. Standing in the doorway was a nubile young thing who asked me if I wanted company. Memories of The Castle Hotel in Mombasa came flooding back. I politely declined her kind offer, mumbling something about having just arrived, and staggered back to bed.

Chapter 13

My first posting was to Flemington, New Jersey, a very rural part of the world. Flemington was a dreamland or nightmare, depending on if you were a farmer or a twenty-five-year-old tea taster fresh from the wilds of Africa. It was quiet and peaceful, bucolic and deadly boring.

I lived opposite an apple orchard. As I arrived in spring, I was able to watch the entire cycle from blossom to apple to take place. I struck up an acquaintance with the farmer, who invited me to help harvest the crop for free apples and a gallon of very hard cider. He shared with me that he was taking a cruise next winter to look for a wife. It seemed the fate of rural men and the search for a spouse were the same the world over.

My life in Africa had revolved around sports and clubs and playing rugby or field hockey or training. It was a shock to find that few played sports in America once they left school. I took to running the country lanes to stay in shape. This was in 1971, before running became popular, and my neighbors regarded me with affection as the nutty Englishman who ran around at night.

On three occasions a police car stopped me as I was running. I had to explain I was not running from something or to something; I was just running. I never did get to understand the social mores of Flemington.

Work was entertaining, though. I reported to the assistant plant manager, who took an instant dislike to me. My combination of accent, youth, expertise, longish hair, and favored status was too much for him. I once irked him to anger when I said that I was "underpaid, oversexed, and over here." He did not see the humor in my remarks.

Everyone at the plant had heard about the trouble it took to get me into the country, I had notoriety before I stepped over the threshold, and he was determined to show me who was boss. His

The modern tea factory is a model of efficiency, but lots of things can go wrong.

name was Falcon, and a more earthbound individual I could not hope to find. His management style was direct. "Do this, do that. You Brits are all the same—can't do a thing without American technology and leadership." We managed by giving each other a wide berth.

The factory was quite modern, with the manufacturing and storage all at the ground level. The warehouse where I kept my teas was to the left, the blending room and tea machines were in the center, and the soup mixing plant and the warehouse for finished goods were off to the right. There were thirty-six tea bag machines; the room adjacent to the tea bag room held the blending drums. The tasting room and all the administration offices were on the second level.

The entire place was run by about two hundred people on a two-shift operation. I reported to the stoic Mr. Falcon, who reported to the plant manager. The latter sat in a magnificent office, chain-smoked, and had long lunches. I was occasionally invited to provide light comic relief. I think they got a kick out of having their own English tea taster on staff.

My job was to taste all the incoming teas against the Lipton standard and decide where they fit in the Lipton blend. My Brash Brothers experience really helped. I would then design a blend using the inventory so that it was an exact match for the standard. The final test to see if I had succeeded in getting it right was to do a mixed batch. This involved taking two cups of the new blend and three cups of the standard, brewing them, numbering the bottoms of the cups one through five, and then mixing them up. If I could not pick out the different blends, then it was a match. If I could tell the difference, I would have to go back and make more adjustments.

I tasted about seventy cups of tea a day. I was busy but not overly so. The rest of the time, I supervised the blending crew. This consisted of talking to the foreman, finding out what was going on, and letting him handle any problems. Mainly these consisted of people not showing up and the crew having to run short-handed.

Each day, hundreds of tea bags would be rejected from the factory floor for various problems. Not having a staple in the bag to close it, having two staples in the bag, not having enough tea in the bag—the reasons were endless. These rejects were then sent through a reclaim machine. This consisted of a cutter to slit the tea bags open and a long tube with holes in it, through which the tea would fall through the holes to be reclaimed. I would then add 5 percent back into the blend on every blend sheet.

The machine was fondly called a ball buster. I was familiar with the term because on the estates in Africa, tea would clump up as it was being oxidized, and we would break up the balls to prevent uneven fermentation. Once, when we had a problem with the machine in Flemington, I wrote a memo describing how a ball buster machine must be serviced. The memo was displayed on the notice board for all to see.

The plant manager called me into his office and told me that the term "ball buster" was just a nickname for the machine. Its official name was "tea bag reclaim machine." Apparently my prominently displayed memo had upset the ladies who worked on the production floor. They had complained about the terminology. I took pains to describe to the management that there really was a machine used in the tea trade that broke balls, but I am not sure they ever believed me. After this incident, my nickname became "Ball Buster."

The plant engineer was a very large gentleman by the name of Lou Slaby. He had hands like dinner plates and shoulders to match. He walked with a lurch as if he had lost his balance; he also had a tight crew cut and looked like the absolute stereotype of an American male. Lou had played middle linebacker for the New York Giants football team. I asked him if he ever played rugby; he said he had heard of it, but it was too rough for him.

For a brief moment Uganda came back into my life when I saw a picture in the local newspaper. By this time Idi Amin had taken over the country, and four local white businessmen were shown lifting Idi Amin in a throne, like a Roman emperor. I knew three of the men and wondered what on earth they were doing messing around with a maniac like Amin. Africa had enough troubles trying to create a sense of nationalism in a tribal society, while at the same time nurturing democracy and the rights of the common man. I counted myself lucky to be far away from that madhouse.

The Flemington plant also made soup. This was a problem. Tea is very susceptible to odors, and the soup mix had a heavy

Chapter 13

garlic content. Tea should never be placed next to anything that is smelly, spicy, or pungent. If not correctly protected, tea will absorb the characters of its neighbor within minutes. I complained about this a lot.

A year after my arrival, the plant manager, Jim Greenwood, announced that tea operations were being suspended and that I should look for a job. I put an ad in a tea trade magazine—"have spoon, will travel"—and I got a couple of offers. But Lipton decided they had too much invested in me. I was offered a position in Galveston, Texas, as a quality control manager.

"Take it, your visa is expiring," Jim said pointedly.

"Okay," I wisely replied.

Chapter 14
Texas Wisdom

GALVESTON IS A STEAMY PORT TOWN ON THE GULF COAST THAT TIME HAS PASSED BY. AT ONE TIME IT WAS THE PRIMARY PORT FOR HOUSTON AND BEYOND. COMPETITION BETWEEN GALVESTON AND HOUSTON LED TO A CHANNEL BEING DREDGED SO THAT LARGE SHIPS COULD SAFELY SAIL UP TO HOUSTON; GALVESTON WITHERED.

Nowadays, Galveston is a city of seedy elegance, run by old money and populated by an interesting mix of poor people, medical students, rich doctors, and a business community that struggles for an identity.

The largest employer was the University of Texas Medical Branch. Situated at the north end of the island, the facility was huge. Half a mile away sat Lipton's plant, a six-story building nestled up against pilings and lapping water. On one side of the road was the plant, on the other a series of bars. It was a rundown street that looked dangerous even in broad daylight.

Old tea packing factories were always built right next to the water so that teas could be unloaded directly into the plant. Each tea chest weights 110 pounds—not an easy item to maneuver. Furthermore, a tea chest is secured by razor-sharp metal edgings.

Chapter 14

These can slice open a hand or a knee in a second. Handling chests took care and big cargo hooks.

The Lipton Galveston plant had one freight elevator that carried chests to the top of the building. Here they were stored and later opened. The tea was then blended and fed by gravity into the hoppers that supplied the machines on the second floor.

Galveston was not an ideal place to have a tea plant. Tea is hygroscopic, easily picking up moisture along with contaminants and other smells. Galveston is a very humid port, and the warehouses were not air-conditioned. In its normal state, tea will stabilize at around 5 percent to 6 percent moisture content. But if it is stored for long periods in a humid climate, it will start to deteriorate.

A four-month inventory of tea was normally available for blending, so it was very important to use the oldest tea first and the latest arrival last: a "First In, First Out" approach, or FIFO.

If packaged correctly after blending, tea can last up to two years. However, if old tea is used in the blend, then the shelf life is dramatically reduced. It normally took at least four to five weeks to cross any ocean to the shores of America. Add this to the month it took to bring the teas from the estate to the port and the two weeks or so to wait for a vessel. When the tea arrived in America, it would be at least three months old. In the trade, this was termed a fresh tea.

Nowadays this time has been shortened. The advent of containers and shorter inventory positions and the use of "Just In Time" (JIT) production techniques have helped. The old way of building up inventories to ensure a steady supply of teas was expensive. The tea cost money and then took up storage space. Introducing containerized tea that blenders know will arrive in good condition has allowed the tea sitting on the ship in transit to be considered as inventory. This has cut down on the amount of tea in storage and means that teas are now fresher when used. It also saves money because you do not have to finance an unnecessarily large inventory. The containers arrive "just in time" for use. It is a real win from the perspective of quality and cost. But tea is still an industry that depends upon overseas sources for its raw material and requires extensive transportation.

My tasks at the tea plant were to manage sixty women who worked two shifts. These ladies measured, weighed, calculated, and adjusted the tea bag machines. It was vital that the correct amount of tea went into every bag. These women ensured that the product leaving the plant was perfect. My mission was to lead them. I was twenty-six years old.

The weather in Galveston was anything but mild, and inside the plant it was quite warm. All the women wore uniforms of a white translucent polyester material; you could clearly see most of their undergarments. With the younger women, this was quite a lure, and they always seemed to be bending or stretching. But maybe it was just my imagination.

The plant was a hotbed of sexual activity. I found Texas women to be a breed apart. They always wear makeup, they always look good, and they consider themselves completely equal to men, just in a different way. For someone raised in Africa, it was an amazing awakening.

I had two elderly assistants, Verna and Lorraine. They guarded the door of my long, narrow, green-painted office. I sat behind them at a gray steel desk and thought of ways to save Lipton money and keep the weight controls in check.

I learnt to calculate the root square mean deviation of a machine. This is a way of determining how consistently a machine can place the same amount of tea into a tea bag hour after hour.

Chapter 14

The rugby team in Galveston, Texas. I'm seated behind the man with the ball.

We weighed one hundred consecutive bags from a machine. We then did a calculation that established an upper control line, above which a bag weight should never stray. We then calculated the lower control line, which was equally unacceptable to cross. In the middle was placed the target weight. The bags were weighed every fifteen minutes.

The tea bag is a relatively recent development in the history of tea; it was introduced around 1904. Tea merchant Thomas Sullivan of New York shipped his tea samples in muslin bags, and customers just dropped them in the cup and poured on the water. Pretty soon someone got the idea that this was a good way for the public to drink tea, and the first tea bag machine was invented.

Before the tea bag, all tea was sold loose. The range of teas available to the public was huge. Any tea grower could send his tea to the grocer's shop and have it displayed in open chests.

Consumers could take half a pound or a quarter-pound of tea or even make up their own blends if they wished—a little bit of this, a little bit of that. There were no restrictions on the size of the leaf, the country that the tea came from, or the taste. Oolongs and large leafy Indian teas or tightly wound Ceylon teas could sit alongside Chinese steam teas with inch-long leaves, all in complete harmony.

Consumers entered tea shops to smell the rich ambience of tea leaves. They could browse for the exotic and leave with a selection of teas to brew in a teapot. The steeping ritual required a pot, a tea strainer, a tea cozy, time, and patience. But the results were worth the wait and worth the trouble. Water would be boiled, the pot warmed.

Then along came the tea bag, and the world changed rapidly. Tea imports dropped off dramatically, not because people were drinking less tea but because they were drinking measured amounts of tea.

With loose tea, the consumer would dig into the tea caddy and ladle out whatever he wanted. One teaspoon per person and "one for the pot" was the norm. But once the tea bag appeared, the extra "one for the pot" disappeared. The pot itself disappeared, and people started brewing their tea bags in mugs.

Convenience overtook tradition. Now everything can be }purchased in neat little paper sachets. However, not all teas will fit into those neat little packages. All the beautiful, long-leaved, aromatic, exotic leaf teas have vanished from the shelves. And with those teas, a lot of romance has disappeared as well.

Size is important, and for a tea bag and its contents, size is essential. Size isn't all that's important, though; there is a concept called "free flow density." In simple terms, free flow density means how easily a product will flow into a small space. Take,

Chapter 14

for example, sugar and cornflakes. Sugar will flow swiftly and easily, whereas cornflakes will get jammed up and not flow at all.

When one deals with tea bag machines, this concept becomes important. Most of the machines work the same way: a hopper holding the blend is positioned above the machine, and the tea flows down a tube. A wheel containing little pockets rotates against this tube. The tea flows into these pockets and is whisked around ninety degrees. The pocket is emptied onto an extended roll of tea bag paper in what is called a "dosage." Depending on the type of machine, the paper is then folded, glued, or covered by a second layer of paper and cut, and bingo. You have your tea bag.

Tea leaves the size of sugar work really well in tea bag machines; cornflake-sized leaves do not. With the switch to bagged teas, the tea world saw a tremendous increase in the use of CTC manufacture, which produces a lot of smaller-sized tea, and a drop-off in the production and use of larger-leaf or orthodox teas.

Lipton saw the market change and quickly switched to tea bags. Soon they established two tea bag packing plants near major ports—Galveston, Texas, and Norfolk, Virginia.

In Galveston, the tea bag machines sat in two parallel lines, spinning and clanking away and spitting out bag after bag. "Printing money" is what the plant manager called it.

Right at the end of the two lines sat the loose-leaf machine. This machine continued to fill the one-pound, half-a-pound, and quarter-pound boxes of loose tea that Lipton still sold. It was a testament to tradition and style but not to efficiency.

Verna and Lorraine kept me informed of all the goings-on at the plant. As manager of the department, I was supposed to make the decisions. But make no mistake, Verna and Lorraine ran the place—and me.

Verna would welcome me with "Good morning, Barry," and a long pause. Then, "Sally will be in to see you today. She needs time off. She has been fooling around with the mechanic who looks after her machines, and her husband found out. He threw her out, and she needs to find a place to live. She can't go live with the mechanic 'cause he's married. He's worried his wife will find out he's been fooling around, so he's dumped Sally. Sally is really upset with him. She is messing with her machine adjustments so he has to come over to fix them. When he does she gives him a hard time." All of this wisdom would be dispended in Verna's lovely, slow Texas drawl.

> ## LOCAL LORE
>
> "Soccer is a gentlemen's game played by hooligans; rugby is a hooligan's game played by gentlemen." Picture a full-contact crossbreed of American soccer and American football, take away any protective equipment, and you've got rugby. Some claim that removing the heavy padded gear makes the sport safer, but either way you won't find many wimps playing rugby.

Lorraine would then chip in, "She needs to find someplace to live. Lord help her, she's looking at the same apartment block the mechanic lives in. So this problem is not going away anytime soon." This was delivered in Lorraine's sharper tone.

On this occasion I went out, and I checked the weight sheets for Sally's machine. Sure enough, the tea bag weights were all over the place. I looked over and saw that the mechanic was just then adjusting the dosage wheels. Sally, a petite blonde, was standing next to him talking intently into his right ear. He looked very unhappy. I decided to steer clear of what I could not fix and went back to my office.

Chapter 14

Actually, managing the women turned out to be okay. Lucky for me, they thought I was cute. "Here comes that cute Englishman," I would hear them say. Verna and Lorraine quickly determined that I needed protecting. They set up a screen around me that proved very effective. Soon the girls and I got along famously.

After a couple of months, Verna and Lorraine just forgot I was in the office. They resumed their normal chatting. I became privy to all the gossip on who was doing what to whom. It was just a very interesting job.

My life was further enhanced when I discovered that the University of Texas Medical Branch at Galveston had a rugby team. They needed a fly half, a role I gladly filled. Rugby being the game that it is, I began appearing at work on Mondays with various cracked brows, torn muscles, scrapes, bruises, and other ailments. The plant manager, Clarence Sutton, a large man who started as a mechanic and had risen through the ranks, called me down to his office.

"What do you do every weekend to get so beat up? Do you lose every fight you pick?" he asked.

"No, I play rugby," I explained.

"Never heard of it," he said.

"Kind of a rough game," I explained.

"It doesn't look good for a member of our management team to participate in something where you get the crap beat out you every weekend. You know that, Barry, right?"

I sort of half-nodded, not prepared to commit myself.

"How come you're always getting injured?" He asked. "You've had a band-aid over your eye for three weeks now."

"Ah, but last week it was the other eye," I explained. Clarence just looked at me blankly.

"People think you're nuts." I could tell he agreed with them.

It turned out that my days at the plant were numbered anyway. I received a phone call from Toby. The familiar gruff voice asked me if I had seen the condition of the teas coming into the plant. I confirmed that I had seen the condition of the chests, and it was terrible. Many arrived cracked and broken, their contents missing.

"Our losses are appalling," Toby said. "We have had to self-insure. We are considering getting someone to look into the problem. I thought you might be the person for the job. Are you interested?"

You bet I was.

"Come up to headquarters next week, and we can discuss things," said Toby.

Chapter 15
Tea Chest Is a Piece of Shit

Transporting tea has always been a challenge. Tea first reached Russia from China on the backs of camels. The tea had been compressed into solid bricks and lashed into bundles. The bricks then spent months bouncing up and down against the sweaty undersides of camels. I cannot imagine how the tea tasted when it arrived.

When Britain and the Netherlands—two seafaring nations—acquired the tea habit, it was a sure thing that ships would enter the picture. Soon the tea clippers emerged—fast, graceful, and packed with tea. It is important to remember that the hulls of these clipper ships were jammed with tea chests into a solid, immovable mass. As the ship pounded through the waves, the tea did not shift at all because it was so tightly wedged in.

From this early experience, a misplaced sense of confidence had been passed down through the ages. It held that that shipping tea packed in tea chests in oceangoing vessels was a thing the mariners knew a lot about. Maybe that was true in the old days, but by 1973, they had forgotten everything they had ever learned. I decided to ask two mariners to find out what they did know.

Tea Chest Is a Piece of Shit

No sooner did I put down the phone with Toby than I picked it up again to call two rugby friends—K. B. Moore (Ken) and P. Kelly (Phil). Both were graduates of the Texas Maritime Academy and were seagoing merchant marine officers. They both knew a lot about shipping tea.

After our next rugby practice, the three of us met in a local pub.

"Tell me about shipping tea," I said, buying the first round of beers.

K. B. was a Texan. He had a Stetson hat that he pulled down low across his brow; he wore cowboy boots and loved guns. He was a man of few words. Phil was from Chicago and was a man of many words. They made a good combination.

"Ship much tea on your ships?" I asked my two rugby chums.

"Yeah. A lot of it ends up in the scuppers, though," Phil said.

"Scuppers?" I inquired.

"Bilges," responded Ken. "Bottom of the ship."

"How does it end up there?"

"Gets crushed." Ken really was a man of few words. Luckily Phil was more verbose.

"Tea is one of the first things loaded aboard the ship," he said. "It comes from the most remote ports we call at. It gets put in the lower hold. Then, when we call at the next port, stuff gets put on top of it. When the ship gets into heavy weather, the tea chests crack, and the tea spills out. After that the chest collapses, then the one underneath cracks, and so on and so forth."

"That doesn't sound good."

"No, it's not a pretty sight. And if we load cargo when it's raining, the tea gets wet, and that's when it ends up in the bilges."

"How can you stop it?" I asked.

Chapter 15

> **TEA TALES**
>
> Even in the 18th century when tea chests were much sturdier they still had to be handled very carefully in transport to avoid damage. Textiles, porcelain, furniture, and other goods were imported from China simply to pack around the crates to prevent damage.

"Simple," said Phil. "Insist on 'tween-deck stowage. That means the tea will go into stow between the first and second decks where the height is about nine feet. After that you cannot stow anything on top of it. Don't ship anything in the lower holds—they are thirty feet deep." He paused. "And strengthen the strapping on the pallets. The strapping is too skimpy, and it comes apart too easily. Also, use cargo netting to hold the pallets in place so they cannot slide around."

"Can we do that?" I asked.

"Sure, you're the customer. All you have to do is tell them where you want the cargo stowed. Tell the ship's agent it's delicate cargo and you want it stowed 'tween decks with netting."

"Containers," muttered Ken.

"Containers?" said Phil and I in unison.

"Put the tea in container ships," said Ken.

"Can't do that," said Phil. "All the tea ships going out to the Far East are break-bulk cargo ships with derricks. The only container ships are going across the Atlantic."

"Stow them on the hatch coaming," said Ken, pulling his hat even lower over his eyes.

Phil looked at Ken with respect. "That's smart," he said.

"What did he say?" I asked Phil.

"He said stow the containers on the hatch covers. The hatch cover fits over the hold. You can stack containers on the hatch covers. You have to take them off when you load the break-bulk cargo and then put them back on before leaving port. That way the tea is in steel boxes throughout the voyage and will never get damaged."

"Tea chest is a piece of shit," intoned Ken.

"What does he mean?" I asked Phil, who had become my translator.

"He means that the tea chest is weak and damages easily. It used to be much stronger. Now it crushes really easily." Phil paused. "Tell you what, come on board the *Stella Lykes*; she's in harbor right now, and we can take you around and show you what we mean."

And that is that is how I became an expert on shipping tea.

I did some background research on the tea chest and discovered it was originally a well-constructed, heavy-timbered, deep box. It was lined with lead to keep out the moisture and then with tissue paper to keep the lead from contaminating the character of the tea.

Over the centuries, plank sides had been replaced with plywood, heavy nails replaced with staples, the twelve corner supports reduced to eight. The tea chest was no longer able to withstand the rigors of modern ocean transport. It was a complex piece of history that time had passed by. It had become a poor container for tea, a terrible tax on the environment, an ecological disaster, and a supply nightmare.

I found that to construct one tea chest required six large panels of 3/8-inch plywood, two yards of foil, and two yards of tissue paper to protect the tea from the foil. It needed sixteen separate lengths of metal edging to fasten the chest together. One half-pound of nails was needed to fasten the metal edging to the chest. Finally, eight batten supports were required to keep the chest upright. There used to be twelve battens, but as a cost-saving measure, it was reduced to eight. All of these components had to be sourced

Chapter 15

Tea chests awaiting transport. They were an economic and ecological nightmare.

and transported to the estate and stored. A perfect supply balance had to be kept among of all the components. Find yourself with all the correct wood but no battens, and you were in trouble.

The tea chest was handcrafted. Every nail, every edging was placed and fastened by hand. The components had to be perfect. If the wood of the battens was too green and had not been cured, then the tea would pick up a terrible cheesy taste. Tasters would immediately spit the tea out. Then the entire invoice—perhaps as many as sixty chests—would be suspect and sold at a loss.

If water had crept in through the edgings, mold would form along the inside of the chest. The taster would detect the rich, earthy taste of mushrooms. Very desirable in a soup, but not in your tea. When an enterprising estate manager in Ceylon discovered a soft drink manufacturer who discarded bad print runs of cans, we

Tea Chest Is a Piece of Shit

saw a rush of tea chests with Coke and 7-Up edgings. The chests were very colorful, but they were weak and fell apart.

If the glue that held the plywood panels in place was rancid, that would affect the taste of the tea as well. Battens that were not cured correctly picked up borer beetle infestations. These little beetles burrowed their way into the wood and happily came out to play once the nice, warm, sweet-smelling tea had been poured into the chest. When the tea-blending manager discovered the beetles, it meant instant disposal and fumigation.

The modern tea chest was an appalling carrier for tea. It was not airtight; it was not waterproof; it cracked and collapsed under the pressure of sea movement. All in all, the tea chest was a disaster, but that was not the worst of it. The worst part of the tea chest saga occurred when the chests arrived at their final destination. Once they had been emptied of their contents, we had to dispose of them.

The Lipton plant in Galveston produced two thousand empty chests a week. We crushed them up and sent them to the dump. They used to be placed in the incinerator and burnt to supply energy, but that created pollution and was made illegal. So the tea chest was crushed into tiny pieces and thrown away. I decided it had to go!

The next week I flew to headquarters in Englewood Cliffs, New Jersey, armed with pictures I had taken on the *Stella Lykes*. I had forty little wooden blocks, representing tea chests, and my newfound knowledge.

I met with Toby, and we both went upstairs to the executive corridor. Here the carpets were plush and the wood was dark. The atmosphere was hushed, rather like a church. I smelled the reverence to money.

Chapter 15

We were escorted into the office belonging to the vice president for material supply and joined a meeting in progress. About ten people were gathered around a table, and papers were spread everywhere. There was tension in the air, and voices were raised. Toby stepped into the fray. I hung back.

I listened. It was clear that the people to the right represented the insurance buyers for the company. They were getting hammered by the people on the left, who were from the operations group.

"How can we run an operation when we plan an inventory receipt of two thousand chests of tea and we get just over half that in usable shape. Tell me that!" one man complained.

"Our losses are running in the millions of dollars. We are almost to the point of self-insuring because no one will insure us." This from an insurance guru in an escalating tone.

There was a momentary silence, and Toby thrust me into the fray. "Allow me to introduce Barry Cooper; he may have some answers for us." Ten pairs of eyes swiveled to look at what Toby had dragged in.

Visions of Ken rose in my mind. I had seconds to grab and hold their attention. I had my story, I had my show-and-tell pieces; what I didn't have at that moment was a voice. Unless I spoke, they would return to the table, and I would become irrelevant, too junior to put forward a view. It was now or never.

"'Tween decks," I said. My voice had taken on a deep, guttural note, brought about by my vocal chords suddenly freezing from a combination of nerves and the spotlight.

"What did he say?" The vice president of operations asked Toby.

Tea Chest Is a Piece of Shit

"'Tween decks," replied Toby. "It's a nautical expression; it means the space between the upper and lower decks." I silently blessed Toby's naval background.

"What about it?" The vice president looked at me.

"Store the tea there and not in the lower hold. Lower hold is thirty feet deep. Stuff gets stored on top of the tea. Tea gets crushed. 'Tween decks is nine feet high. No stowage on top." I had picked up the speaking pattern of my hero, Ken.

There was silence. I filled it.

"Use nets to stop the pallets sliding around and banging into each other. Ship containers on the hatch coaming covers; introduce containers into tea ports." I hesitated, and then dived in.

"Tea chest is a piece of shit."

"That's enough, Barry. Thank you." Toby raised a hand, but I had the vice president's attention. He waved me forward.

I pulled out my first set of two blocks of wood. I had duplicated a tea pallet. The "tea chests" were positioned in four layers of five interlocking chests each. They were held together with three rubber bands. One band secured the top layer of chests horizontally. The other two bands were stretched vertically over and around the blocks.

"This is the current strapping configuration on all the pallets that we ship from overseas." I handed the model to the vice president. "Shake it around a bit," I said. "Imagine it on a pitching ship in heavy seas."

He did, and within seconds it fell apart. I handed him the second set of blocks. On this set I had secured every layer with a rubber band horizontally and four sets vertically.

"Do the same with this one," I suggested.

He did, but it did not come apart. "The cost of a pallet of tea is about two thousand dollars," I said. The vice president looked

Chapter 15

over at the insurance guru, who nodded. "The cost of that extra strapping is six dollars," I told them.

"Tell me more about 'tween decks and nettings and containers," the vice president said.

Twenty minutes later, I had a new job: manager of Raw Tea, Ocean Transport, and Inspection, or RTOTI, as it soon became known. I was to be based in Galveston because, as a port, it was the place to see the results of the improvement. I was to report to Toby because he could understand what the hell I was talking about. I was to travel to all the overseas ports and countries from which Lipton purchased tea. I was to make sure they used netted, 'tween deck stowage and introduced containers. Oh, and by the way, I was to get rid of that shitty tea chest.

It was an amazing opportunity for a twenty-eight-year-old. The largest tea company in the world had just given me a mandate to mess with tradition and change the way tea was shipped, something that hadn't been attempted in hundreds of years. I had a clear objective and the support of senior management because they all wanted results, and I could travel to the remotest parts of the world to fix the problem. I was shaking with excitement. What a job!

Toby and I had lunch before I flew back to Galveston.

"Well done in there, Barry. I was a little concerned at first. Your speech patterns seemed to have changed, but you soon made yourself comfortable and carried the day. How did you learn so much about shipping tea? Spend much time messing around in boats, do you?"

I flew back to Galveston in a daze, and a week later I was in Sri Lanka, introducing the concept to the Lipton offices in Colombo. Claude Godwin was the general manager there. He immediately

Tea Chest Is a Piece of Shit

understood the need and the idea behind the program. His target was the tea chest.

"The tea chest is a useless anachronism. Should have been done away with years ago," he said. "In the old days, the chests were made of planks of wood and had twelve battens. Now they are made of eight battens and have plywood sides. They are not strong enough to withstand the journey. They crack open at the riveting weld down the side of the chest."

We visited the Ceylon Standards Institute. These were the folks who told the tea estates how to construct the tea chests and what to use in the construction. Because they were a government agency, their word was law. We had to convince them that the tea chest was a piece of shit. Here the director patiently explained to us that the Ceylon tea chest was perfectly okay. It was designed to withstand a drop of three feet on its corner and not split asunder. We were assured we were wasting our time. The tea chest was a fine container for tea.

Claude arranged for a test to take place. The institute's management showed up, pristine in starched white shirts and immaculately pressed flannel trousers. The eight-batten chest hung suspended three feet above the concrete, its corner aimed at the floor. We measured the distance to assure ourselves it was three feet.

"Ready? Okay. Let go!" The laborers released the rope and then jumped back as the tea chest hit the floor and exploded in a billow of dust. The institute's management team remained frozen in place. A swirling brown cloud of microparticles enveloped them. The dust settled, and they stood there, their world in tatters.

Claude raised his arms. "It's got to go, you know. Its time is past. It is an ecological nightmare, and it doesn't work," he said. No one moved.

Chapter 15

Ten days later I met in London with a manufacturer of natural kraft multi-wall paper sacks. I wanted to find out what we could do to make a sack that could hold one hundred pounds of tea and withstand shipments from distant climes. The demise of the tea chest had begun; the date was 1974. Toby and I were about to change the tea world forever.

Chapter 16
Flight Out of a War Zone

THE PLANTERS LOVED THE TEA CHEST. THEY WERE USED TO IT AND WANTED TO KEEP IT. THE SUGGESTION TO REPLACE THE CHEST WITH A FOIL-LINED PAPER BAG, STACKED ON A PALLET AND WRAPPED IN PLASTIC, WAS MET WITH ASTONISHMENT AND THEN LAUGHTER. CLEARLY TOBY AND I WERE NUTS. BUT WE ALSO REPRESENTED THE LARGEST TEA-BUYING COMPANY IN THE WORLD, SO THE LAUGHTER WAS SHORT-LIVED.

It was not only polite, it was good business sense to listen to what we had to say. But as far as they were concerned, the tea chest left the estate in perfect condition. They never saw the condition of the chest on arrival; that was the problem. The only one who saw the pictures of the damage was the insurance company.

I began sending the planters pictures of the tea when it arrived in Galveston. Slowly the estates began to understand the need to change. We developed a system for shipping it in Lighter Aboard Ship (LASH) vessels. Just one barge could deliver two thousand chests of tea all at one time, with no losses to the Galveston plant.

Chapter 16

We introduced ventilated containers. This allowed the tea a chance to breathe and prevented excessive moisture buildup during transit. We developed a multi-wall, foil-lined tea sack. It was adopted as an international standard and accepted by the tea trade as an approved method of shipping tea. Most tea is now shipped in containers, packed in bags on pallets. This has become the normal way to transport tea. We had changed the tea world forever!

But changing the tea world was not without risk or adventure. I was traveling to remote ports and tea estates. On a trip to Mozambique to introduce this new way of shipping tea to a group of planters, I became trapped in the middle of a revolution. Mozambique borders the Indian Ocean and runs down the East Coast of Africa. It was colonized by the Portuguese, who were not benevolent masters, but they had established a vibrant tea industry that supplied the world with good, plain medium teas. Mozambique also had two very good ports at Ncala and Beira. These ports were also used by landlocked Malawi in Central Africa. Making sure that Ncala and Beira could handle the new tea sacks was important, so I headed there to oversee the job personally.

I was the only occupant on the twin-engine Fokker flight from Nairobi to Beira. The main port was shut down. The shipping agents who were supposed to meet with me did not appear at the hotel as scheduled. All the hotel staff had also disappeared. After two days of waiting for someone to contact me, I decided it was time for me to go as well.

Independence to Mozambique had been granted by the Portuguese because a revolutionary movement called FRELIMO had risen up and started shooting people. In undue haste, the Portuguese handed over the country to an elected government that the Portuguese supported and crossed their fingers that the government would stay in power. It didn't. At the time of my arrival

Flight Out of a War Zone

Tea chests routinely got smashed during transit, ruining their contents.

in Beira, the FRELIMO guerillas were still shooting people. It was unfortunate timing, as they didn't seem to mind whom they shot. It that sense they were a democratic movement—they shot at everything, so danger to everybody was equal.

Amid the sporadic sound of gunfire, I found a taxi driver who was prepared to take me to the airport. We encountered three roadblocks and safely negotiated them, but we arrived to find the airport closed. All flights had been suspended. I returned to the cab, but the driver had suspended operations as well and left. There I was, suitcase in hand, a revolution going on, no flights, and no place to go. I went back to the terminal and saw, tucked away in the corner, a counter with a large sign announcing "Charter Flights." Behind the counter a man was staring at me intently.

Chapter 16

> ### LOCAL LORE
>
> Mozambique was ruled by the Portuguese dictator António de Oliveira Salaza, and during his rule, there was a massive influx of Portuguese immigrants. The push for independence began in 1962 with the formation of FRELIMO, the Liberation Front of Mozambique. FRELIMO took control in April of 1974, and within a year almost all of the Portuguese colonists had left, either expelled by the government or fleeing in fear. Unfortunately, they took most of the infrastructure with them when they left, and Mozambique remains largely undeveloped.

"Are you flying charters?" I inquired.

"Where you want to go?" replied the man with the intense gaze.

"Anywhere."

"We go there."

"How much?"

"How much you got?"

We negotiated a price, and Captain Gonzales outlined the plan.

I was to go to the bathroom and stay there for half an hour. At the stroke of noon, I was to leave and walk outside to the car park nearest the terminal. When I saw a light, single-engine, high-wing plane taxi past, I was to jump over the low fence, throw my suitcase in the plane, get in myself, and we would leave. No customs, no immigration, no exchange control—we'd just leave. He wanted half the payment up front, half on arrival.

I moved to the toilet, found a stall, and sat there quietly for half an hour. My mind was in turmoil. Should I leave the country without exit papers? What would happen to me if I was caught jumping over the fence? What if they shot at me? What if they shot at the plane?

Despite the risk of leaving a country illegally, I did not see many alternatives. I had no idea how long this revolution was going to last. I did not speak Portuguese. I was a white face in a black country, so it would not be easy to hide. The plane seemed the best of a lot of high-risk alternatives. I tried to remember if I had seen any fighter jets as I landed.

After half an hour of low-grade panic, I left the stall, only to meet an armed soldier walking in. My panic level skyrocketed. He politely stood aside and watched me struggle out of the stall with my suitcase. I washed my hands and left. He was probably wondering what the only passenger in the entire airport was doing in a toilet stall with his suitcase. Maybe he had other more pressing things on his mind because he let me go on my way.

The fence was low and my leap high as the plane approached. Within a minute I was secured in the co-pilot's seat. Captain Gonzales turned onto the main runway and revved up, and we took off.

"No control tower," he yelled over the engines as we went wheels up. "No flight plan. You got insurance?" With no flight plan, there was no record of our departure. If anything happened to us en route, no one would know.

We stayed low over the jungle, then slowly edged up until we were high in the clouds. Captain Gonzales relaxed, put the automatic pilot on, and told me we were heading for Rhodesia.

The year was 1974, and the Unilateral Declaration of Independence affirmed by Ian Smith against the United Kingdom was in effect in Rhodesia. As the holder of a British passport, I might not receive a friendly greeting, but at least they were not shooting at each other.

We droned on until suddenly another plane appeared alongside and checked us out. Captain Gonzales waved, and minutes later

Chapter 16

we made our approach into Salisbury Airport. We were arrested the moment we got out of the plane and marched to an immigration room. I showed my passport. Captain Gonzales appealed for political asylum.

I was allowed to go free after half an hour, a piece of paper stamped "Entry Permit" stapled to my passport. They kept Captain Gonzales, but he seemed relaxed and happy to receive the balance of his payment.

Two hours later, I caught a South African Airways flight to Johannesburg; I was soon back in the civilized world. I had the foresight to ask Captain Gonzales for a receipt for my money and to warn Toby to expect a slightly unusual expense report.

What he got was a single sheet of paper, no heading, with just the handwritten words, "Received from Barry Cooper, the sum of 750 U.S. dollars for a charter flight out of a war zone," dated and signed by Captain Rudolpho Gonzales. It was never questioned.

There are two postscripts to my Mozambique adventure. The first occurred in the mid-1980s when I was invited to go back to the country to investigate reclaiming a tea property. It was an estate that had produced good tea in the past but was having trouble starting up again. A consulting company had asked me if I wanted to join them in the project.

"What's the problem with the property?" I asked. There was a slight hesitation before an answer was provided.

"Landmines."

"Landmines?"

"Yes. You see, when the rebels were fighting the government, they were indiscriminate about where they hid landmines. No record was kept, and now people and animals are being blown up all over the place."

"And you want me to go look at this place?" I asked in bewildered amazement.

"Well, we asked around, and we've been told that you're an adventurous type."

"Not that adventurous."

The landmines issue destroyed the Mozambique tea business. What used to be a thriving industry exporting in excess of thirty thousand metric tons of tea a year now exports nothing. People and animals are still being blown up. No one has come up with a solution to the problem.

The second postscript occurred in a Paris café, thirty-one years later. I had ordered a coffee, and the waiter commented on my bush jacket and accent.

"From South Africa?" he asked.

"No, from Kenya," I replied.

"Ah so. My family came from Mozambique. We left in '74," he responded.

I mentioned that I had been in Mozambique in '74 as well and began to tell him the story of my departure. Mid-tale he stopped me and clapped me on the shoulder. His eyes were alight with laughter, and a broad grin was on his face.

"I know the ending, sir," he exclaimed. "It was my cousin who flew you out. You are a piece of my family history."

He went on to tell me that Gonzales lived in South Africa and that he still flew. I passed on my regards. The coffee and baguette were free.

Chapter 17
Diving into Specialty Teas

MY WORK AND TRAVELS FOR LIPTON WERE PAYING OFF. WITHIN THREE YEARS, WE REDUCED THE LOSSES TO ACCEPTABLE LEVELS AND EVERYBODY WAS HAPPY.

I was promoted to Lipton headquarters in Englewood Cliffs, New Jersey. Headquarters was a large flat-topped building just off the Palisade Parkway, minutes from the George Washington Bridge. I was made assistant to the director of tea buying. I was placed in control of all the tea buying for Lipton instant tea. I was also to manage the RTOTI function.

After traveling the world, sitting in an office buying tea was not very exciting. It irked my soul. When a friend called to tell me a mail-order specialty tea and coffee company was for sale in New York, I was intrigued.

In the world of tea, there are two very distinct businesses. The major companies—like Lipton, Tetley, and Brooke Bond—sell teas that are better than average, but they are not specialty teas. Specialty tea companies concentrate on very small volumes of rare and expensive teas. Such teas are sold through mail order, and today, through the Internet.

Such companies cannot afford large inventories. Nor, very often, are large quantities of specialty teas available. Instead, suppliers

Diving into Specialty Teas

work with wholesalers who do buy large lots and then subdivide them into smaller packs of one to five kilos of tea. These are then further subdivided and sold to the public. Simpson and Vail was such a company.

Simpson and Vail was located at 53 Park Place in New York City. The owner had died, and the company was part of his estate. He had been sick for a long while, and the business had declined dramatically. I could buy it cheap.

A friend arranged the loan I needed to buy the shares to control the company. I was now deeply in debt. All the money I had built up in my years in America was on the line, and failure was not an option. But with my knowledge of the different tea estates and my experience venturing into remote areas, I was ideally suited to such a business; I was more than happy to travel and seek out rare teas from around the world. There were six specific countries to focus on—India, China, Formosa, Ceylon, Japan, and Kenya.

INDIA

Chapter 18
Countries of Origin: India

INDIA IS A LAND OF SMELLS—PUNGENT SPICES, DECAYING VEGETATION, SMOKE FROM VILLAGE FIRES, INDUSTRIAL SMOG, POLLUTED RIVERS, POOR SEWAGE SYSTEMS. THE SMELL STAYS WITH YOU LONG AFTER YOU HAVE LEFT THE CONTINENT. LIKE A MEMORY IT PERMEATES YOUR BRAIN, NEVER TO BE FORGOTTEN AND IMMEDIATELY RECOGNIZED THE MOMENT YOU SET FOOT IN INDIA AGAIN. IT IS A LAND THAT EITHER DISTURBS YOU OR ONE THAT YOU LOVE.

The poverty of India is overwhelming. It is a country of teeming cities, of crowded trains tearing down the track with passengers clinging to the roof carriage like ants, of glorious buildings of the Raj, now dust-stained and grimy.

My family history is intertwined with India. My grandfather was in the Royal Horse Artillery. He commanded troops in the Khyber Pass in the extreme northwest of India. He also commanded the Red Fort in New Delhi. My father was born in Peshawar, now part of Pakistan; my aunt in Calcutta; my uncles in Bombay. But India was a hard life for a soldiers and wives, and it was worse for young sons who needed to be educated. At the age of five, my father was

Chapter 18

> **TEA TALES**
>
> Many people have found their fortunes in tea, but it was a man named Fortune who made it all possible. Born in Scotland in 1812, Robert Fortune traveled to China for the English East India Company and managed to smuggle out around twenty thousand tea plants, which he brought to Darjeeling, India. This broke the Chinese monopoly on tea and established the tea market for Europe. Fortune was also the first European to discover that black tea and green tea are made from the same plant.

sent back to England to go to school.

I grew up on stories of India. My grandmother's home in England was cluttered with memorabilia of India and the Raj. Tales were oft told of lone Englishmen commanding native troops under dire circumstances. And it was the heroic efforts of an intrepid Englishman—one Robert Fortune—that pierced the Chinese veil of mystique and introduced tea to India.

Disguised as a laborer, Fortune, under penalty of death, smuggled thousands of tea plants out of China. He even brought with him a gang of tea makers. With their coveted knowledge of how to turn the two little leaves and a bud into salable tea, Fortune created the Indian tea industry.

Fortune worked for the English East India Company, better known then as "The Company." The Company established tea gardens in India, thus becoming the first tea producer free of Chinese control. But tea in India was under the control of the British, which many would now say was just as bad.

The British soon introduced new tea processing systems. They mechanized the handmade Chinese traditional teas and increased production dramatically. They also discovered that the local Assam tea bush was actually a better tea producer than the Chinese types. Plus it grew all year round. This was wonderful news; tea was popular and in fashion, and the Company strove to meet the demand. With their new Assam source and systems, they were soon doing so.

In the tea trade it is said that you are either an Indian man or an African man. There is no doubt as to where my loyalty lies; it is with Africa, that much is surely clear. But the rolling tea fields of India captured me, as did those of other countries. The hot, steaming valley of the Brahmaputra River fascinated me. Most good-quality tea is grown at high elevations, yet in Assam the tea is grown at sea level. It has established its own unique place in the world.

Assam tea that is grown in the fall—or, as the trade would call it, an "autumnal"—has a richness and malty character that distinguishes it from its compatriots in a single sip. The taste is coupled with a deep rich color, highly valued by blenders.

The Assam leaf is broad and a deep green with slightly serrated edges. It can be twice as large as a China tea leaf, so it is easy to spot in a tea garden. In Assam, the bushes are densely packed. Moisture is not a problem. The taproots of the older bushes do not have to go very deep to reach the water table.

Revolutionaries have made the Assam Valley a dangerous place to visit today. The United Liberation Front of Assam (ULFA) wants independence for the region. They claim the Indian government illegally occupied land in 1947 when India became independent. It a Maoist organization that operates with little pity. Kidnapping and ransom payments fund the group, which has been brutally effective in slowing down development of the region.

The guerillas intimidate the civilian population, who have little to gain by standing against them. When the Indian Army tries to

Chapter 18

pursue, the terrorists melt away into the jungle. It is a conflict with little hope of a resolution. It has been going on for so long that the tea trade has just come to live with it, though visiting the area remains a high-risk undertaking.

I did go, and I was given an armed escort to make me feel more secure. Lunch at one estate consisted of me, the estate manager, three armed guards, and some sandwiches. But most of the terrorists were running around with AK-47s. My guards had bolt-action, single-shot .303 Lee Enfields. They knew they were at a severe disadvantage. They seemed to spend a lot of their time on the extreme edge of my vision, ready to leave at the first site of trouble, and I couldn't blame them. Luckily, their weapons were never put to the test.

Darjeeling, on the other hand, welcomes visitors. Darjeeling nestles in the foothills of the Himalayas in Northern India, at elevations of between five thousand and seven thousand feet. On a clear day you can see Mount Everest soaring over the world, pristine in a blanket of snow. The tea bushes crowd in on the town. You grind your way up into the mountains on the famous spiral railway, which dips and curves like a roller coaster. The train even goes backward at times to gain sufficient momentum to get up the steep inclines. You can see bushes clinging to life on hardscrabble ground; tight groups of those bushes sprout from rocky outcrops on steep hills. The narrow, sharp leaves are tell-tale signs of their Chinese origin. They struggle against the elements and the high elevation. The first time I made the trip, it was scary going up; as you climb the mountain, you are exposed to a panoramic vista of green hills and a terrifying drop straight down. I began to think about warped rails and loose ties and the boiler going bang. Though it seems inconceivable, the road runs alongside the railway. It is a perilous route populated by fume-belching buses,

Countries of Origin: India

In many ways India is still a country of old.

motorbikes, haphazard taxis, tea-laden trucks, and intrepid private citizens in their cars.

My grandmother had described the railway trip to me when I was a boy. Back in the early 1900s, she rode the train to escape the summer heat of Bombay. I imagined what it must have been like for my grandmother. In those days, as the wife of a senior officer, she would have had her own carriage, perhaps even two—one for her, and one for the nanny and the children. They would start the journey in the cool of the early morning and escape the heat with a steady rhythmic climb up the mountain.

As I inspected the polished glossy wood and the yellowed sliding glass windows of the train, I felt connected to my long-dead grandmother. She must have been in her late twenties, around my age, when she made the journey. Her first trip up to Darjeeling must have been a wondrous adventure for her, and I was thrilled to follow in her footsteps fifty-five years later. Every turn and twist in

Chapter 18

India is also home to a rapidly growing high-tech industry.

the tracks opened up a vista, unchanged from what she must have seen in her youth.

My return trip from Darjeeling was by car. Going down the road at night was terrifying—on the one side a steep drop into nothingness, on the other the railway line. The turns in the road were so abrupt that the headlights of the car swept out into the black night, and there was no reflection. It was just night sky and space.

Regardless of the difficulty in reaching Darjeeling, a tea taster feels the journey is worth the trouble. The tea areas are spectacular; deep valleys set against the backdrop of the Himalayas provide awesome scenery. But growing tea in Darjeeling is difficult. The gardens are old and require a lot of care, investment, and extensive fertilizing. Nor is the tea easy to pick, as it clings to very steep slopes, and the bushes are sparse.

A good example of a specialty item is the first growth of spring tea called first flush. In China, India, and Japan during winter, the tea bush stops producing buds. For three or four months, the tea bush is at rest. When spring sunshine revives the bushes, tiny succulent buds burst forth. They are gorgeous to look at and tasty to drink. The leaves are translucent and wafer thin, yet they possess a strong, pungent character and flavor. Darjeeling first-flush tea is a standard for the specialty tea business.

Many first-flush teas are thin in the cup. This means they lack substance and mouth feel. They have a distinct greenish character, and this flavor is much admired among some consumers. When you add milk, the tea turns a pale gold; however, I advise against adding milk to any of the Darjeeling teas because, in my opinion, they are not made for milk. Add a squeeze of lemon if you wish, but why contaminate a tea that has fought so hard to reach your cup?

China

Chapter 19
Countries of Origin: China

NO ONE REALLY KNOWS HOW MUCH TEA THE CHINESE PRODUCE BECAUSE THEY DRINK SO MUCH OF IT THEMSELVES; THE BEST ESTIMATE IS ABOUT NINE HUNDRED THOUSAND METRIC TONS A YEAR. I SUSPECT IT IS A LOT MORE, AS IT IMPOSSIBLE TO KEEP TRACK OF THE TEA MADE AND CONSUMED IN THE VILLAGES THROUGHOUT CHINA. TO PUT THAT PRODUCTION FIGURE INTO PERSPECTIVE, THE UNITED STATES IMPORTS ABOUT ONE HUNDRED THOUSAND METRIC TONS A YEAR.

I first went to China in 1978, just after Henry Kissinger and President Nixon had opened up the doors of diplomacy. China was eager to sell its tea to the world's largest tea company. Immediately after the president had made his trek to China, the Chinese government invited my boss to visit. Toby went, he saw, he came back, and then he sent me. At the time, I was responsible for the movement of tea around the world for Lipton. After visiting the major tea-growing area and seeing the infrastructure of roads and warehouses, it was clear why I had been told to pack my bags and get over there.

Chapter 19

In 1978 everyone in China wore blue Mao jackets. I was accompanied by a man from head office and a local man from the region, and I am sure they were both security people. They spent a lot of time watching each other, having decided that I was completely harmless. These were the days before tourism, and I was conspicuously obvious.

I spent most of my time in the far reaches of the land. What I saw was a huge country bursting with energy and cultivation. Food was grown everywhere—by the roadside, in gardens, in flowerpots. Wherever there was an open space, food was grown. Everybody I met was industrious and disciplined. Over the next ten years, I made frequent visits to China, always to different parts of the country. I came to know it well, and I saw it go through some amazing changes. Those changes continue to this day.

My China stories are at times harrowing. On my third visit, I was flying from Kwangsi to Hunan in a local twin-engine DC-3 prop plane. Suddenly, the plane dipped—gradually at first, then at a steep angle. There was a loud bang, and one of the engines stopped. The pilot made a terse announcement in Chinese that clearly worried all the Chinese passengers; they became very agitated and started a group wail. The pilot then paused to say in broken English, for the benefit of his sole European traveler, "Hold on!" That made this sole European traveler just as agitated as his Chinese companions.

We landed safely, but my guides and traveling companions were clearly shaken from the experience. It was eventually conveyed to me that we would drive to the next tea-growing area. We did, but we found ourselves in a very rural area that evening. We stopped at a compound and were invited to stay for the night. I was treated as an honored guest and served a boiled egg, along with a cup of very good tea, before I retired for the night. The room was a humble one, lit by a kerosene lamp. It had one rope-slung bed, a single blanket, and no pillow. I bundled up my jacket for a pillow, turned down the lamp, and was soon asleep despite the surroundings. You make do in such an adventure.

I awoke to a loud crunching sound. I turned up the kerosene lamp and saw my stash of Kit Kat bars in the process of being eaten by a huge rat. This creature would have put a poodle to shame. I looked at him; he raised his head and looked at me. I began to pray that the Kit Kat bars would be enough for him.

The next day, we started off and were soon in the familiar realm of rolling hills and tea bushes. The Chinese tea bush has a long, thin leaf, unlike the Assam leaf, which is broad and thick. The Chinese leaf is built to survive in cold climates and goes dormant for three months of the year. This period leads to a buoyant spring harvest and much celebration about how good the first flush of spring tea tastes. Though the country is more often associated with black teas for export, China actually grows and consumes mostly green tea.

> ## Tea Tales
>
> The difference between green tea and black is a matter of processing. The same two leaves and a bud that are picked to make green tea are used to make black tea. Green tea is tea that has been picked and immediately heated in some fashion to stop any enzyme reactions. This process keeps the chlorophyll content of the leaf intact, which makes it stay green. There are many different ways to stop the enzyme action. Any heat source will do it. Steam will do it. Boiling water will do it. An oven will do it. A frying pan will do it. All of these methods have been used, and are still used, to make green tea.

Chapter 19

The Chinese have been making tea for thousand of years; it is a part of their daily life. One cannot begin a meeting in China without first going through the ritual of accepting a cup of tea. You watch politely as it is brewed for you. Every office, hotel room, and meeting room has a thermos flask of hot water and a little wooden box of tea. If you are in a tea region, you will find the local tea in the box; otherwise it will be jasmine tea, which is ubiquitous and popular.

The process is always the same. A handful of leaves is dropped into a mug, and hot water is poured over the leaves. A neat ceramic cap is placed over the mug to retain the heat. You are then handed your mug, and you remove the cap whenever you feel the time is right. It is polite to murmur "ahhhh" to get the meeting off to the right start.

Because of this ritual, I wanted to learn more about jasmine tea. Most jasmine tea you find in the West is green tea with jasmine flavor sprayed onto the leaves. Jasmine is so expensive that most of the flavor is artificial. You can tell real jasmine tea; dried jasmine petals nestle among the finished tea leaves. I wanted to learn how the Chinese could create such wonderful jasmine teas by using only real jasmine flowers. How did they get the aroma and flavor without adding oil?

I asked the Chinese to show me their production techniques. This caused consternation—some methods the Chinese like to keep secret. After much nodding and gesturing, they told me that they would show me how jasmine tea was made and that I was an "honored guest," deserving of such a courtesy. I translated this to mean that if "big buyer" wants to see jasmine tea being made, we are going to show him. At four a.m. the next morning, I was awakened by a pounding on my bedroom door. "We go see jasmine," they told me.

After two hours of a jarring car ride, we arrived just as dawn was breaking over a commune with rolling green bushes. It looked just like a tea field. Closer examination showed that it was a field of jasmine bushes. Each bush contained tight little bundles of jasmine blossoms, closed against the cool of the night but ready to spring open as the sun's rays warmed them. Only they never got the chance.

The field was full of young women. They were bundled up against the cold as they went from bush to bush, plucking the blossoms and placing them in wicker baskets slung over their hips. A soft fragrance hovered over the scene. A gentle mist was rising. Dawn was breaking. It was too romantic to be real.

We then followed the wicker baskets on their journey. Workers piled them into a truck and drove it over to the next valley. There I saw the familiar rows of tea bushes set among a series of flat-roofed buildings, which looked rather like cattle shelters, only longer and more slender. In these buildings, I discovered men with wooden shovels spreading dried tea evenly throughout the hut to a depth of about a foot.

They added the tightly closed jasmine blossoms to the tea in an even layer. The workers then spread a second layer of tea over the blossoms. I examined the big, flat-sided wooden shovels. They looked exactly like the snow shovel that I had at home and achieved the same purpose. They moved large amounts of material the old-fashioned way, by hand.

The Chinese then took me to a second shed and revealed the mystery of how jasmine tea gets and retains its flavor. The jasmine buds, picked at dawn while they are still closed, generate heat when they are covered by the tea. This is because they are moist, and moist vegetative matter creates heat when covered by dry matter.

Chapter 19

The saying "all the tea in China" still carries a lot of weight—it's a lot of tea!

Any farmer who has lost a barn to a fire caused by stacking wet hay under dry hay will tell you that.

This heat causes the buds to open. The freed petals then release their fragrance, which is hungrily absorbed by the tea. This whole process is repeated until the tea has retained a distinct jasmine character.

There are different grades of jasmine, depending on how many incubations of buds are introduced to the tea. To achieve a really fine jasmine, the ritual is repeated six times, giving the tea the name sixfold jasmine. Such a tea rarely finds its way to the West. It is too expensive; it is too delicate. It is savored and consumed in China on special occasions; if you are lucky, it is given to you as a gift. I was lucky, and I treasured my jasmine for a number of years.

Lapsang souchong is another tea that intrigued me. Lapsang souchongs are very smoky, almost tarry, teas. They have a distinctive taste and character. In many ways, the taste is similar to the Scotch whiskey called Laphroaig, which I also enjoy. I asked to see how this tea was made. Of course, I had read in tea books how it is smoked over a peat fire, but I wanted to see the process myself. How long was it smoked? How much was smoked at one time? These are the kinds of details that a professional really likes to comprehend. Learning it from a book isn't quite the same as seeing it take place.

The Chinese told me, "it is not convenient," and that was that. That statement, in China, means "cannot do." If you push the issue, you embarrass your hosts. It is a phrase you learn to acknowledge quickly and politely.

Chinese tea is grown in a number of very distinct provinces: Kwangsi, Yunnan, Hunan, Hainan, Hubei, Jiangsu, Fujian, Sichuan, Keemun, and Zhejiang. The people of every province are of the opinion that their tea is the best; it is like tracking down wines in France. In the area where the grapes are grown, people always think their wine is superior. And the truth is, when you taste the freshly made tea among the bushes where it was grown, it is the best tea you have tasted. Just as wine tastes superb when it is tasted among the trellises upon which the grapes have hung.

On a later visit to China, I did get to see how the lapsang souchong was made. Long, spiraled leaf teas were placed in smoking huts; each hut had a round hole in the pitched roof. The hut was filled with tea, except for a space in the center where a peat fire was made. Smoke soon filled the hut and slowly made its way out through the roof hole in gentle wafts. In the meantime, the tea absorbed the peat character. I am not sure that what I saw was the traditional method that produces the fine teas I have tasted in the past called lapsangs. The whole setup looked to me like a very haphazard affair. It did not look like a commercial enterprise;

Chapter 19

> ## Tea Tales
>
> Longjing (meaning "Dragon Well") tea is a famous variety of green tea grown in the Zhejiang province in China. The name was derived from the brewing process, which used water from a well containing dense water. After a heavy rain, the lighter water on the surface would exhibit a sinuous, twisty boundary with the water below it, and the movement was thought to resemble a Chinese dragon. It is believed that the best Longjing tea requires this special water to be boiled and then cooled to around 80°C before being used for brewing. Genuine Longjing tea is expensive—selling for upwards of seventy-five dollars per pound—and because of the tea's price, the market abounds with fakes.

rather, it appeared to produce a tea made for the local market and taste.

A tea I have never liked is Pu-erh, a deliberately aged tea that can be stored for years and tastes musty and old. It is used as a medicine in China. The Chinese introduced me to the tea in Quandong, and ever curious for a new tea taste, I dived right in. Big mistake. It was foul-tasting, moldy, and bad. I spat the brew out with venom, and my hosts were polite and looked away as I coughed and spluttered through the batch of ancient Pu-erh teas they had prepared. They explained it was good for me. I took their word for it but have avoided Pu-erh ever since.

Another fine specialty tea from China is Dragon Well; it is one of the finest green teas in the world. It is revered for its character, taste, and aroma. It can be found only around a certain well named the Well of the Dragon, which is located in Hangzhou in the coastal Zhejiang province, just south of Shanghai. This is one of the oldest teas in China. It was an inspired tea taster who figured out that the tea's extraordinary jade appearance and its taste is actually due to the fresh spring water that feeds the well and the nearby tea bushes. To this day I hear comments about how superb Dragon Well tea is. It is superb, but it was the water that made it famous.

White tea, or silver tips as it is sometimes called, is another specialty tea found in China. It is the rarest of all teas, a single bud that is picked by hand and dried without the aid of any processing. Only certain tea bushes produce buds of the correct length and spiral to mature into white tea. China has a white tea that the emperor Hwei Sung (1101–1126 AD) considered his exclusive domain; he would execute any trespassers found in the tea fields during spring harvest.

While I owned Simpson and Vail, I purchased cartons of white tea and wrote in my newsletter about the rarity of such a tea. One article explained the "Agony of the Tea Leaves," which describes the unfolding of the buds as hot water is applied. In fact, white tea does even more than that—white tea actually dances. As water heats the buds, air trapped in the curled leaves expands and causes the buds to rise. Because each leaf is a different size, the amount of air trapped is different, so the leaves rise in a sequence. As they reach the surface, the air escapes, causing the bud to sink back to the bottom, passing its compatriots on their way up.

Watching white tea steep is a mesmerizing experience, unique in the world of tea. My customers at Simpson and Vail loved the story and the romance, and my white tea offerings sold out. I ordered more, but the supply was exhausted for the season, and I had to wait until the next year. This is often the tale in specialty tea; there is not much grown, which is what makes it special.

Chapter 20
Countries of Origin: Formosa

I VISITED FORMOSA—OR TAIWAN, AS IT IS NOW CALLED—AND WAS AMAZED BY THE ANTIQUITY OF THE EQUIPMENT. WHEN GENERALISSIMO CHIANG KAI-SHEK ESCAPED THE COMMUNISTS AND MOVED TO THE ISLAND, HE BROUGHT WITH HIM MASSES OF TEA-PROCESSING EQUIPMENT. MOST OF IT WAS STILL OPERATING WHEN I TOURED THE ESTATES.

On the wall of my office I have a picture. It shows a man with a long pigtail sitting high in the seat of a tea sorter. At the base of the machine, another coolie is collecting leaf into different baskets. The picture was taken sometime around the 1860s.

Imagine my astonishment when I found the same equipment operating with the same efficiency when I visited in the early 1980s. All that was changed was that the pigtail had gone. I walked around the wooden frame of the tea sorter and searched for a manufacturer's plate, but it no longer existed. The machine, in all its splintered, clattering grandeur, continued its task, more than a hundred years after it was designed. It was magnificent, but it wasn't commercial.

The art of tea making is dying on the island of Formosa. I found a willing and eager audience in Simpson and Vail customers, who savored the fine and expensive points of Formosa tea. Formosa is famed for oolongs. These are semi-fermented teas that have a smoky, grassy character. At the lower level of quality, the taste can be coarse and unpleasant, but the better teas are exquisite—light in color, delicate on the tongue, with a warm, toasty, roasted, nutty aroma.

Formosa has become an industrialized country, and tea production has dropped dramatically. High wages have attracted farm workers to the cities, so there is less labor to produce the teas for which Formosa was once famed. Small quantities of very fine oolong teas are still produced with care and at great expense. These teas are consumed locally or are exported to Japan. Very occasionally, these teas are exported to Europe. But Formosa tea is a dying industry. It will never disappear—tea is too much a part of the Chinese way of life—but as a force in the tea world, it is spent. Formosa tea should be enjoyed while it lasts.

There are those purists who will be upset with that last paragraph, but when was the last time they visited Formosa? I have been to Formosa. I have watched the tea being made. I have seen it rolled by hand, with methods developed hundreds of years ago. It is quaint. It is original. It is fascinating. It is wonderful to watch. The teas taste superb. However, it is totally impractical.

In an age of superlatives, hand-rolled and hand-tied teas are another epitome of extravagance, curiosity pieces that cannot meet any sort of demand. In Formosa, to hand-roll teas, the withered leaves are first placed in a cotton bag with a drawstring. This is tightened until the ball is malleable but firm. Then the worker

Chapter 20

> ### Tea Tales
>
> Oolong (meaning "black dragon") tea is a Chinese tea that lies somewhere between green and black in terms of oxidation. Legend has it that a tea farmer was frightened away from his drying tea leaves by a black serpent. When he returned several days later, the leaves had oxidized under the sun and created the oolong type known today.

begins to massage the tea back and forth, as if rolling out a cigarette.

This produces great style, as the leaves are twisted into elongated shapes. The rich juices that are pummeled out of the buds and leaves slowly stain the cotton bag. After half an hour of exhausting work, the cotton drawstring is released. Over half the tea is then removed and fired immediately in a small oven.

The drawstring is then tightened down into an even smaller ball, and the process begins again. I watched the arm muscles of the worker flex and bulge as he slowly created the second batch of tea from the bag. At this point, the bag was soaked with the tea juices and was discarded. A fresh one was selected and a new batch of tea introduced into the cotton pouch.

The entire process had taken over an hour, and I estimated that less than a pound of tea had been made. The industry is struggling to stay alive; the costs are too high. Simple economics will decide the fate of the Formosa tea trade.

Hand-tied teas are even more unique. Withered leaves, still flexible and easy to turn, are knotted into a variety of shapes and designs. Women with long slender fingers complete this work. I have not seen the process, but I have seen the end result. It is extraordinary; the teas look wonderful unfurling in the cup, and their shapes are works of art.

Perhaps that is the best way to look at such teas—as works of art. They are to be appreciated for the talent of their creators, but they are not to be considered of this world. They are of a different realm of tea—that of the curious, not the real.

❧

CEYLON

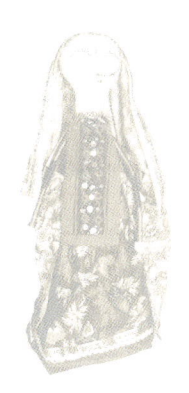

Chapter 21
Countries of Origin: Ceylon

I FIRST SAW CEYLON WHEN I WAS FIVE YEARS OLD, WHEN THE SHIP CARRYING ME, MY MOTHER, AND MY SISTER TO MALAYSIA STOPPED IN COLOMBO. I REMEMBER SWIMMING IN A POOL THAT WAS FED BY SEAWATER AND BEING SURROUNDED BY THOUSANDS OF BRIGHTLY COLORED FISH, A KALEIDOSCOPE OF WAVERING COLORS. MY NEXT VISIT WAS TWENTY-FIVE YEARS LATER—THE POOL HAD LONG SINCE VANISHED BUT NOT THE HEAT, THE HUMIDITY, OR THE BUSTLING INTENSITY OF COLOMBO.

Ceylon, or Sri Lanka as it is now known, is one of the largest producers of orthodox teas. Ceylon is a jewel of an island that sits at the tip of India; it exports over three hundred thousand metric tons of tea a year—about twice as much as India, its huge neighbor. In the world of tea, Ceylon is a giant.

Ceylon is an island of reaching peaks and deep valleys. The highlands extend down the center of the island in a huge ridge

Chapter 21

that blocks the monsoon winds and creates a microclimate that is ideal for growing tea.

The greenery of Ceylon is overwhelming. Luscious tropical plants loom over every doorway and roadway. Exotically colored plants, thick, twisting vines, and tall, soaring trees dominate the countryside. It is also a land of stupor and traffic mayhem. Cars, brightly ornamental trucks, and dilapidated buses gushing noxious fumes stumble and roar all over the place in a cacophony of blaring horns.

In most tea-producing countries, the original planters—be they Dutch, English, or French—left a wonderful collection of buildings in the cities they settled, and Colombo is no exception. To walk on the seawall, in front of the old parliament building, is to step back in time to a slower, more elegant era.

Ceylon is blessed with two monsoons a year—the southwest monsoon, which runs from June to August, and the northeast monsoon, which lasts from December to March. This weather pattern makes the island unique in the tea world, as it has two seasons for producing high-quality tea.

High-quality tea is grown when the weather is warm, the mornings are cool, and the rain is infrequent. This weather forces the root systems to produce flushes of leaf that are intense and flavorful, particularly at the higher elevations. At the same time on the other side of the mountains, the tea fields receiving the monsoon rains get drenched. They produce a lot of leaves, but the teas are very plain and are called "rains tea."

The gift of dual monsoons originally led to the start of a coffee industry in the early eighteenth century in Ceylon, but a coffee blight wiped out all the bushes in the early 1860s. Someone suggested growing tea, and the first fields were laid out about ten years later. The idea was brilliant, and tea prospered.

Countries of Origin: Ceylon

Not surprisingly, it was the Scots who opened up the first tea gardens. I think only a race of their stubborn determination would have persevered in the early days. Many Ceylon tea estates bear Scottish names: St. Andrews, Kirkoswald, Craighead—the estate names must have brought back distant memories to those first lonely planters. Homesickness aside, the pioneers carved estates out of virgin jungle and created an industry that dominates the world of tea to this day.

There are three elevations at which tea is grown in Ceylon. Teas grown under two thousand feet are called "low-growns." These have superb style, but their cup quality is not that good. The teas are much sought after by Middle East buyers, who admire the carefully crafted leaf styles.

From two thousand to four thousand feet, medium teas are grown. These teas do not have the pungency or astringency of the high-grown teas, and they do not have the blandness of the low-growns. They are good blending teas with good leaf style. They are of drinkable quality and are reasonably priced. The town of Kandy, situated in the middle of the island, is the center of the region that produces medium-grown teas.

Kandy is pure romance. Temples draped in heavy vines surround a pristine lake. Exotic flowers brandish themselves on every patch of soil, and elephants can be seen wandering the streets. Brightly colored saris wound around slim bodies weave in and out of teeming throngs. Rising above the town are the hills that spiral upwards into the highlands.

Take the road out of Kandy into the highlands, and you enter another world. Above four thousand feet, there is the sound of silence. It is one of the few places in the world that I have found to be really quiet. Not even airplanes can be heard. There is the crackle of silence. Teas grown here are considered high-grown teas; they

Chapter 21

Ceylon still has a an aura of romance.

are prized and priced for their clarity, taste, aroma, and flavor.

The town of Nuwara Eliya is the center for high-grown teas. The town was called a hill station in the old days. It was a place where the planters' wives, daughters, and the occasional girlfriend would go to escape the heat of Colombo. It is reminiscent of rolling English hills. Here it is too cool for tropical plants, so the hills are covered with short grass, which is verdant green and lush. Horses and cattle graze serenely on the pastures, and there is an overall sense of peace. The peaks that surround Nuwara Eliya are full of whimsically named estates: Oliphant, Lovers Leap, and Tommagong. Despite their esoteric names, the estates produce teas of extraordinary depth and quality.

While the journey to Nuwara Eliya may be an adventure, it is always worth the trip. The teas are just that good. Within this area is a group of about six estates where at a certain time in late spring, the teas take on a lemon note. There is no known reason why this takes place—it just does. The production of this tea is too small to interest the big companies. It is also too obscure for most specialty tea companies to know about. But I did, and I garnered a small supply of these exquisite teas, selling them to connoisseurs who enjoyed the story, the romance, and the taste of these rare items.

Each estate is a world unto itself. They have their own hospitals, nurseries, schools, day care centers, community centers, sports fields, housing for workers, churches, and temples. An ancient rhythm pervades life in the hills, bringing one back almost to the time when the earth was lit by fire.

The estate manager rules everything. The estates produce a never-ending stream of tea that must be handled every day. Bushes do not stop growing, so the leaves must be picked to keep the bushes under control. Once they are picked, the teas must be processed. Teas left too long without rolling become over-withered and bitter and will lose value. So, like an army, the estate's organization must be run with discipline and structure. Everyone must know his job and be on time to start the tasks for the day. It is the role the estate manager to instill this discipline and integrity into the organization.

The price his tea fetches is the benchmark of a planter's success or failure. If a tea is poorly made, the value it receives in the Colombo auction will be reduced, and reduced prices cannot be allowed to continue for long. So the planters have stressful lives in this idyllic kingdom.

In the old days, the planting fraternity let off steam with raucous parties, hunting expeditions, and wild exploits. The graveyards of the up-country churches are spotted with tombstones that bear witness to the success of tigers and wild boars in exacting their own revenge. Stories abound of horses that could find their own way home with the master of the house unconscious and draped across the saddle after a wild night of drinking. But that was the past. The pursuits now are more sedate. The paved road to Colombo has made the city accessible to wives and children. Estates are no longer as isolated,

Chapter 21

and there are now many families in residence. Hospitality still rules the day, however, and every weekend sees a round of dinner and cocktail parties where the latest tea prices are compared.

The European planters are a vanished breed. Men of similar desire and fortitude who are native to this land have taken their place. But the new generation reveres the work of those that went before.

Visitors from overseas are particularly welcome. They bring news of the outside world. The world of tea is a small one, so gossip is demanded. Who was doing what, to whom, and where? Who has been promoted? "Whatever happened to so-and-so?" will be leveled at you, and you are expected to know these facts. Then you will be given a curry so hot that the soles of your feet will burn.

There will always be a farewell party to send you on your way. The planters get together and pass along messages for you to carry back to Colombo and beyond. You will be plied with drinks and expected to keep up with the others.

On one such visit with a fellow tea buyer, we scheduled our departure too late. The road down to Colombo is unlit, tight, and twisted. It is not for the faint of heart. But I had a flight to catch the next day, so depart we did.

After half an hour of slow, melodic curves, I felt my head slump forward as sleep approached. I glanced over and saw that my fellow traveler was also in the head-down position, his head bobbing to the bumps in the road. I looked up toward the driver and saw that his head was bobbing as well!

"Driver!" I yelled, causing the car to swerve violently. "You were sleeping."

"Oh, no, Sahib," he said. "I was just saying my prayers. Please do not be alarmed. I know all the places to crash."

Chapter 22
Countries of Origin: Japan

I WAS VERY FORTUNATE TO VISIT JAPAN WHILE I WAS DEEPLY IN LOVE WITH A JAPANESE WOMAN. MY PASSION FOR HER WAS HOPELESS AND HELPLESS. SHE CAUSED ME TO LOOK AT THE BEAUTY OF TREMBLING LEAVES MISTED WITH EARLY MORNING DEW. ONCE I SAT AND WATCHED A ROCK GROW.

I tried to understand the Japanese appreciation of stoicism, beauty, and Zen-like grace—all of this to impress my love. Her name was Mariko, and she was the heroine of James Clavell's novel *Shōgun*. Her grace and courage shone through the book, and when she departed the novel, courtesy of a foul plot and a large explosion, I was a lost soul for a while.

Tea is a part of the fabric of life in Japan, as it is in China—even more so, perhaps, as Japan has developed a tea ceremony that is as much an act of tranquility as it is of brewing tea. In fact, the tea brewed in the tea ceremony does not taste very pleasant at all. But for those of us who have participated in the ritual, the ceremony puts the participant in such a Zen-like state that everything is good. One small complaint is that the tea tends to leave a film of grit clinging to your teeth, but a moment of Zen is worth a slightly itchy feeling in your mouth.

Chapter 22

> ### Tea Tales
>
> The Japanese tea ceremony, or *chanoyu*, is renowned as the height of the art of tea drinking due to its elaborate rituals and required knowledge not only of tea but also of kimonos, flower arranging, ceramics, and incense. It can take years to learn and a lifetime to master. The most basic tea ceremony requires a variety of different tools, including the pot and cup, whisks, special shelves, ladles, and other items, all of which are handled with the utmost care and sometimes require special gloves. The objects are often works of art themselves.

In *Shōgun*, Mariko is told by her lord and master to make peace with her husband. She does so by inviting him to a tea ceremony, where civility and tranquility must reign. It was one of the most fascinating features of Japanese life, this appearance of peace and tranquility. It actually masks a visceral violence; at least, that is my view. The politeness on the surface is all about the suppressed emotions underneath. Sometimes that emotion is not too suppressed, either—all it takes is a few drinks, and the façade is allowed to slip. Dealings then can be very direct and open.

During the day, the Japanese complimented my wisdom and skill as a tea taster, and I felt I had impressed my hosts. In the evening, after a couple of beers and glasses of Suntory whiskey, they told me that I could stay in Japan for the rest of my life and still not be able to taste all the varieties of teas that they made. But in the same breath, they said they would do the best they could for me. They felt I had a certain talent—undeveloped, perhaps, but worth spending a little time and energy on.

Chanoyu, the Japanese green tea ceremony, can truly take a lifetime to master. In principle, it is all very simple. They reverently place a fine powdered green tea in a bowl and pour hot water over it. They then beat the brew with a bamboo whisk into froth and allow it to cool slightly, and the tea is then sipped. The devil, of course, is in the details.

The angle the wrist must assume while beating the tea, the number of beats it must receive, the number of wands to the brush—all of these elements are part of the lore of the ceremony. The list of requirements is long, and the ceremony is an event that must take place in harmony. Those who attend and participate must be at peace with themselves and the world.

The event is held in a *roji* (tea garden), where the *jo-an* (tea room) is situated—away from noise and confusion, and always near water. The sequence of how the utensils are laid out is prescribed; even the method and sequence of entering, sitting, and standing are all dictated by tradition and form.

The tea caddy, or *chaki*, can hold the two types of tea used in the ceremony. The first tea, called *usucha*, is a thin tea made from bushes up to fifteen years in the ground. The other tea, called *koicha*, is made from tea bushes that are at least twenty years old. The latter brew is pasty, resembling a thick, green, gritty soup. It does, however, have a stunning aroma, full of the heavy green notes of cabbage.

The rhythm and grace of an experienced hostess performing the tea ceremony is part of the magic. I could understand how the ritual has gained the importance it has to the Japanese way of life. The ceremony is an oasis of discipline and quiet. It is soothing to listen to hissing coals, gurgling water, the rustle of silk kimonos, and the whisk of bamboo on porcelain while smelling the tea aromas. It is mesmerizing. If only Mariko had been there to enjoy it with me.

Chapter 22

Tea first came to Japan from China by the ubiquitous Buddhist monks in the twelfth century—or in the ninth century, depending on whom you believe. The seeds were first planted around the ancient capital of Kyoto. The Japanese initially used classical Chinese methods of cultivation; however, over time, the Japanese moved away from frying their tea in pans to stop the oxidization and introduced steam drying. This process retained the color of the tea and introduced the classical vegetative taste that fine Japanese teas maintain to this day.

There are some interesting grades of Japanese tea. Bancha consists of roasted tea stalks and is drunk as an iced tea in the summer. Sencha is another common grade, but don't be fooled—there are over nineteen grades of it. I say over nineteen because I have never been able to get a straight answer from anyone as to how many there really are; I am not sure they even know themselves. There are low-grade senchas and very high grade senchas. The difference is similar to the difference between generic jug burgundy and a Lafite Rothschild vintage—they look the same, but that's about the only similarity. To my Simpson and Vail customers, I offered the most rare of the sencha teas.

The premium tea, and probably one of the world's most famous, is the Japanese gyokuro. This tea is made from bushes grown in the shade. It has a deep green color and is delicate and distinctive in the cup. It does not possess the vegetative taste of the senchas.

The deep green color of gyokuro is achieved in a typical Japanese fashion. As soon as the first flush is blossoming, workers cover the tea fields that are to produce gyokuro with matting. This covering prevents light from reaching the tea bushes, allowing the tea leaves to retain a higher chlorophyll level and thus a much darker green color when the tea is plucked and made. As with the rare white tea from China, gyokuro is produced only in May during the spring harvest.

Over the years, labor has become so expensive in Japan that they have had to introduce mechanical plucking. One system that has been developed looks a lot like a hedge trimmer attached to a billowing net bag. The workers hold the hedge trimmer in two hands and sweep it across and over the contoured bush. The tender tea leaves are pushed back into the bag. Less selective machines are used for the coarser sencha teas.

There are also harvesting machines that take two men to operate. These devices clip the tea bushes in a nonselective fashion, chomping lots of stalk. Once the tea is made, it has to be filtered and sifted through a variety of screens to remove the stalk.

There are three plucking seasons in Japan. The first season, which begins in late April and ends in May, is the spring season. The second flush season begins in late June and runs into the final season, which occurs in early August. Japan can be subject to brutally cold winters and sudden spells of frost, so many of the larger

> ## Tea Tales
>
> Gyokuro—which means "precious dewdrop"—is considered the highest grade of tea available in Japan. The tea bushes are grown in shaded areas so that the leaves produce more chlorophyll, making the final tea greener than other types. It also can use a slightly different brewing process than other green teas. Generally, you will use slightly more dry weight than with other teas for the same brew. Brewing takes place at cooler temperatures, around 50°C–55°C, and a brew time of around two to three minutes. Since gyokuro steeps at such a mild temperature, it is customary to heat the pot and cup first so the tea stays warm while being drunk.

Chapter 22

Automated harvester machines have been introduced to reduce labor costs.

estates have windmills and fans that move the air around during the cold periods and stop the bushes from freezing. As in China, the winter months are fallow, and no tea production takes place.

For all its tradition, Japan has been quick to adopt modern tea concepts. It is the largest market in the world for ready-to-drink (RTD) teas—the type of tea packaged in bottles. In the United States, these teas often contain very little tea. Not so in Japan. Most Japanese RTD teas have powerful tea notes and character. They are not for the faint-hearted.

Tea estates around the world use Japanese steaming techniques to provide green tea to Japan in the Japanese style. The large Japanese community in Brazil is a major provider of tea to Japan, as is Ceylon. Black teas are sold in Japan, but only the very finest grades are acceptable. Japan is a unique market; like its people, it is complex, bound by tradition, and sometimes difficult to understand.

While there, I learnt to nod politely, drink my drink, and try to see the funny side of things. At one memorable dinner, my hosts asked me to choose my meal from a pond behind the bar. I watched the fish slowly swimming in circles and pointed out a bright yellow and black specimen with elongated fins. The Japanese congratulated me on my choice and told me it was "good eating."

The victim was netted and shown to me; then it disappeared behind a swinging bamboo curtain to meet its fate. Half an hour later it reappeared on a platter. Wafer-thin pieces, artistically arranged around a brightly colored fish.

How clever, I thought. The chef has painted the fish. I was reaching with my chopsticks for the first raw sliver when the fish suddenly flipped up, flapped its tail, and mouthed a few gasps.

It was alive! Not only that, but it was surrounded by itself, sliced into neat segments. I felt like a cannibal. The fish had been filleted, but on one side only, thus leaving all its intestines intact and functioning. It gave a new meaning to the word *fresh*.

KENYA

Chapter 23
Countries of Origin: Kenya

"Are you married, or do you live in Kenya?" This was something you often heard as I was growing up in Africa. Kenya had a roguish reputation, partly brought about by it being the dumping ground for ne'er-do-wells and unproductive second and third sons of the aristocracy. These folks reveled in the club, farm, and family life in Kenya—although most of the time it was other families that they were reveling in.

This belief in Kenya's depravity was confirmed by the murder of Josh Errol in a case that was never solved. Everybody knew who had shot Josh—it was the husband of the woman he was having an affair with; it did not take a genius to work that out. But there was a certain laissez-faire attitude about fidelity in Kenya. It was considered amusing to pursue others; it was certainly amusing to be pursued. In the case of Josh Errol, the husband got away with it, and the wife divorced him and married Lord Delamere.

This tradition of infidelity was born in the country homes of England, where it was customary that a junior houseman would sound a gong at four in the morning, allowing those in the wrong

Chapter 23

beds to return to the correct ones. It was a part of country life. After dinner and cards, what else was there to do?

This culture made its way over to Kenya, where the far-flung towns were hotbeds of flirtation and conquest. Molo, Kericho, Nakuru, Eldoret—all had their various reputations. But at the same time, these amoral fly-by-nights were carving massive farms, tea estates, and coffee *shambas* out of the wild bush. Living rough, they were not afraid of hard work; those that stayed in Nairobi acted as commercial agents.

Kenya was the perfect place to grow tea. The Kenya Highlands, at five thousand feet, were cool and lush with rich soil and plenty of rain. They remain that way to this day, although the rampant Kenyan population explosion has meant a lot more villages dot the landscape than did when I was a boy.

The British did a fine job of establishing tea in the territories. Independent planters and huge land-owning corporations both flourished in the free-for-all that was the colonial age. Large tracts of land up-country were reserved for whites only and became known as the "White Highlands." This act of folly created enormous resentment and was immediately dismantled after independence. But up until then, it allowed the colonists to prosper and create the basis for what is now the one of the largest exporters of tea in the world.

As in Ceylon, it was coffee that was grown first. It was not until 1903 that tea was introduced with cuttings brought over from India. Kericho, an area that nestles up against the shores of Lake Victoria and is west of the Rift, became the center of the tea trade in Kenya.

To select Kericho as the first place to grow tea was an act of genius. The proximity of Lake Victoria enabled cool, moist air to from the lake to drench the tea bushes, ensuring their survival and eventual maturity. It was a long way from the coast, however, and to get to Mombasa, the main port for Kenya, required an arduous journey. But this pales in comparison to the difficulties that the early planters in India and Ceylon faced.

And therein lies the major difference between the Kenyan tea industry and those that had gone before. In India and Ceylon, men tore up virgin forests with the help of elephants and with their bare hands. It was brutal and inspiring at the same time. Empires were literally carved out of the jungle.

This was not the case in Kenya, where the industry was established right from the start as a business. The tea market was already a well-established fact by 1903, and the London auctions were in full swing at the London Commercial Salerooms on Mincing Lane. In 1937, they moved to Plantation House, also on Mincing Lane, where I later sat as a trainee. It was here that the buyers and sellers gathered every week to bid on the teas shipped in from the far-flung reaches of the Empire and beyond. The men who created the Kenyan tea industry were visionaries, certainly, but not adventurers. For them it was a business with a very clear objective. Not for them was the life of a lonely tea planter, sleeping rough under the stars, rifle tucked by his side to fend off those that thought of him as dinner. In Kenya the approach was organized, structured, and successful.

My first visit to a Kenyan tea estate occurred when I was nine, and I wish I could remember which one I visited. It could have been the Brooke Bond estate, Mabroukie, but I am not sure. I know it was east of the Rift and just outside Nairobi. We drove up there one Sunday afternoon. What did register was the long, winding drive through the dust-covered tea bushes lining the road. The road branched toward the top of the ridge, dipping to the left

Chapter 23

to the factory and climbing slightly to the right to the manager's house.

It was long single-story home, its roof a thick covering of bundled native grass like a thatched cottage in England. A low veranda encircled the home with views over the factory and the rolling tea gardens. In the backyard was a large vegetable garden. It was green, peaceful, and entrancing.

The interior of the home was all polished rough-hewn planks, comfortable sofas, and chintz-covered lounge chairs. I remember book-lined walls and large dogs. My sister and I were given permission to go and play with the horses, and the groom took us out through the tea bushes to the paddocks.

The air in the highlands of Kenya is tantalizingly clean, and the grass has the rich, thick look of being well-nurtured. Things grow there. Like all tea estates in those days, there was tremendous self-sufficiency. Some were fully functioning farms as well. I know that when we sat down to a laden table at dinner, I could see across the table and over the window ledge to the rolling vista of tea bushes. It struck me, even then, as a good life to live.

Going back to Kenya to visit the estates is always a shock to me. It feels strange. My memories are of dusty roads, long low bungalows, horses and cows, and tea factories with narrow fermentation troughs and wooden floors. Life was slow; tea took eight hours to make.

Now it is all tarmac roads, heavy trucks, and high-speed equipment that hums and spews out a steady streak of macerated leaf in a high arc to automatic fermentation beds in just ninety seconds. Where there used to be a sense of artistry and design, now there is efficiency and stainless steel. The same wonderful smell of freshly cut tea leaves, a cross between a perfume and a green vegetative

COUNTRIES OF ORIGIN: KENYA

Workers in Kenya blending tea the old-fashioned way.

aroma, still gives you that heady high as you walk into the factory, but it is all business and output now, not artistry and guile.

The Mabroukie estate was east of the Rift, as were the Aberdare Mountains, home of Mount Kenya, a grade-A alpine peak that was snow-capped for my entire youth. It is east of the Rift where all the fine high-quality teas are now grown. Kericho is still a major producer of tea, but the tea gardens around Mount Kenya, Limuru, and Kiambu, where I grew up, produce a better-quality tea. I believe the teas have more flavor and character. I know they have captured a very loyal following of buyers from around the world.

When the industry was established, the normal, traditional, orthodox methods of making tea were used. It is the machine equivalent of rolling a withered green leaf in the palms of your hands and breaking it slowly into smaller pieces. This gentle way

Chapter 23

of making tea took up to eight hours to complete and produced some wonderful flavors. The Michimukuru estate, near Meru, was my favorite. Their tea had a bright golden color and a distinctive malty, sweet, penetrating character that I found delicious. As a trainee in London, I would sneak the bowl off the batch tray after the evaluation tasting and smuggle it back to my office to drink; it was that good.

But the new method of making teas—cut, tear, and crush (CTC)—has virtually replaced all the old orthodox equipment. The advantage of the CTC method is speed. The leaf is ripped and shredded in seconds and produces a large percentage of a similarly shaped particle ideal for use in tea bags. It also produces the qualities that Kenyan teas are now famous for.

Because of their strength, color, and character, Kenyan teas have replaced Indian teas in many international brands. The Indians consume most of their own teas, and India's tea exports over the years have actually decreased, leaving them to struggle to maintain their position as a major tea exporter. This little-known fact has given Kenya second place in the world's most important tea-exporting countries, right behind Ceylon.

Kenya also has the most successful smallholders venture in the history of tea. Named the Kenya Tea Development Authority (KTDA), the organization has been responsible for the huge success of tea in Kenya. Small landholders, rather than large plantations, each grow an acre or two of tea and pluck the leaves every ten to seventeen days, depending upon the amount of rainfall they receive. These leaves are then purchased by the KTDA for an annually agreed price. It is a steady cash flow for the farmer and a reliable source of green leaf for the KTDA.

The Authority assists in planting to required specifications and advising on weed control and fertilizer application. It also provides for the management of the bushes with pruning and plucking instructions and making sure the farmer fills in his fields with new bushes should he lose a few to drought or disease. This tea is processed in forty-five different factories and produces one hundred and eighty thousand metric tons of tea a year, more than the entire tea production of Indonesia. The smallholders concept has made Kenya a major force in the world of tea; the Mombasa auction that is held every week on a Monday is a key determining factor for the world price of tea.

Privately held estates still exist, and companies such as Brooke Bonds, the African Highlands Company, and Eastern Produce continue to ply their trade long after the "White Highlands" are just a vague memory. Their affairs are managed by the Kenya Tea Growers Association, which is still based in Kericho.

Some things have stayed the same, and the Kericho Tea Hotel, where my father and I used to play golf, still stands. The Brooke Bond Company built it, and it still looks out over the tea fields. The hotel also still serves a really outstanding afternoon tea. It now plays host mainly to groups and tourists. But if you close your eyes and sit back on the old colonial sofa, teacup in hand, the hotel closes in around you, and peace descends upon your soul. I like to go there. It is a comforting mix of the old and the new, sufficient to allow me to accept the changes that have taken place.

Chapter 24
The Challenge of Selling Specialty Teas

At Simpson and Vail, I learned to be a tea merchant. I packed small amounts of fine teas into packets and sold them at very nice prices. The business concept was simple. High-priced teas, much desired by the connoisseur and not available in the local supermarket, could be purchased from me by mail. To reach my customers, I had to let them know what I had on offer. I did this by sending out a catalog.

I sent out four a year, one every three months. The previous owner had lapsed into sending out a mimeographed sheet just listing the teas. This was easy to do but did nothing to explain or describe the teas or show the leaf style. I upgraded the catalog and added black-and-white pictures and stories about the tea, and I introduced romance into the process.

I had to learn a whole new way of doing business, as the staff members that had stayed with the company were little more than

The Challenge of Selling Specialty Teas

Some of our marketing materials. A shame the phone number was wrong.

order-takers. They manned the phones, and that was it. If I was going to build the business, it was going to be my efforts that made it happen. I found myself negotiating hourly rates with photographers and arguing with the printers over print minimums and the quality and type of paper to be used. Then I would find that the printer had glued the catalog and not stapled it as I had wanted.

"Not my fault," said the printer.

"Don't care about whose fault it is, I just want my catalog the way I ordered it," I said.

"Okay, well, I can reprint in three weeks' time," the printer said.

"Not good enough. I am supposed to mail it next week."

It was an education. The photographer was never on time and always ran over budget. He seemed more interested in what teas I would leave with him than in taking the shots we needed.

Chapter 24

I designed a new logo for the business, which looked great, and had an artist friend draw it up. I asked that the address and phone number of Simpson and Vail be placed directly under it and ordered a print run of thirty thousand, which was delivered in twelve large boxes. The logo looked fabulous, and the address was correct. The problem: the phone number read (201) 344-6377-8. Our area code was 212.

Nothing was easy. Everything required my involvement. It was an aspect I had not considered. I was used to Lipton, where I had secretarial help and physical help when I needed it. At Simpson and Vail, if I wanted something done, there was no question as to who was going to do it—me.

My employees watched my efforts with wry amusement. I think they knew long before I did that I was fighting a losing battle. Unbeknownst to me, my business was in trouble from the moment I acquired the company. The mailing list was out of date by at least five years. More than half the addresses weren't valid. The former owner had been too ill to expand the business, and the list had suffered.

I found this out when more than half of the catalogs I sent out came back marked "return to sender." As that pile grew, I realized that I had purchased a mail-order business that did not have a list of customers that could sustain it, and I was to suffer the consequences. The mailing list is the cash-flow generator for a mail-order business. Without an active list, we were sending catalogs out to the general public hoping to generate a sale, not knowing if the people who got the catalogs even had any interest in teas. It soon became apparent that if the company was going to survive, I was going to have to broaden the offerings.

I delved into the rarest of the rare. Little does the world know that there are niches of the tea world where teas are so unique that the flavors last just a few weeks. The first-flush Darjeelings, the Assams of the new season, and spring silver tips are fine examples of teas with limited production. The idea worked, and I enticed new customers into the fold, but it still was not enough to keep the company afloat.

Simpson and Vail was not going to survive without an injection of capital. I approached a few banks. They all reared back in horror at the suggestion that they lend me money to spend on mailing and catalogs—all "sunk funds," in their words. So I was, indeed, sunk. There was nothing left to do but find a buyer for the company.

The company had a good old name and a good reputation. What it didn't have was an owner who had the money to build up the mail-order list again to make it a functional business. If it could be moved out of the city to a rural setting and perhaps coupled with another company, Simpson and Vail could be a very viable business.

First I had to take care of my friend who had arranged for the loan to buy the company. I did so by signing a note that had me paying back the loan, with interest, over a period of years and handing over my shares, such as they were.

A suitable buyer had to be found, and as luck would have it, one appeared at the right moment. Jim Harron still runs Simpson and Vail. From its rural location in Brookfield, Connecticut, it dispatches fine teas all over the world.

The fact that I had lost all my money in the venture was not as meaningful as the fact that I had learnt a lot about business. I knew I had tried very hard and had done some good things. I also knew that my lack of business experience had led me to buy a company without going into its books and truly understanding its worth and situation. It had been an expensive lesson, with too

Chapter 24

many days of dismay and not enough days of "Eureka!" But on those days when it did all come together, I experienced a sense of freedom, independence, and self that was intoxicating.

※

Chapter 25
Herbs

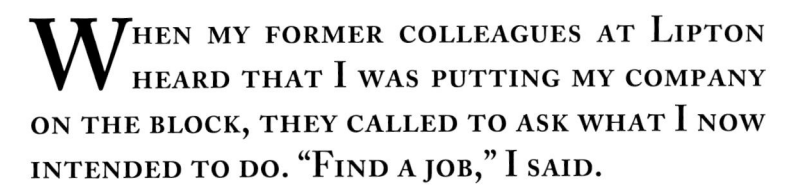

WHEN MY FORMER COLLEAGUES AT LIPTON HEARD THAT I WAS PUTTING MY COMPANY ON THE BLOCK, THEY CALLED TO ASK WHAT I NOW INTENDED TO DO. "FIND A JOB," I SAID.

"We might have one for you," they said. It was music to my ears.

I was to interview for the position of Lipton's herb beverage research and development manager. My interview for the position was with Dr. Hal Graham, vice president of research and development. Hal was a tall, austere man with a great sense of humor. We knew each other from my old days at Lipton. He had once gone on safari in Africa. My mother owned a safari company at the time, and he had used its services. Our meeting was brief; it went well, and I got the job. During the interview, Hal mentioned he was headed to Africa again.

"Is there anything I could take your mother from America?" he asked.

"I'll call her and find out," I said.

"Dahhhling, yes," my mother said. "How very kind of him. I would *love* one of those new padded American toilet seats—*so* comfortable."

I reported Mother's request back to Hal.

Chapter 25

"Ahhh," he said. A kind, courteous man, he had probably thought I would give him a letter to hand-deliver to her.

Later, he recounted his experience at the Kenya customs hall when he landed at Nairobi. Hal walked majestically through the hall, with the seat wrapped in paper and suspended over his arm.

"What is that?" asked the customs inspector, pointing at the seat.

Knowing that gifts would be charged a duty, Hal responded, "It is my toilet seat."

"You travel with your own toilet seat?" asked the skeptical customs inspector.

Hal drew himself up to his full six foot three and replied, "It is padded."

"Let me feel it," the customs officer demanded, and he leaned forward to squeeze the soft contours. "Aaaahhh," he said, giving Hal a nod. "Yes, that would be very comfortable; you must take great pleasure in sitting on such a seat."

The year was 1979, and a small company in Boulder, whimsically named Celestial Seasonings, had made herb teas a category on the shelves of the supermarkets. Celestial's team took the revolutionary step of adding natural flavors to herbs. Overnight, a new beverage category was created. Celestial also had the genius to make their packaging look like herbs; customers could stand in the supermarket aisle and almost taste the products through the graphics. Sales soared, and other tea companies took notice. Lipton wanted a piece of this market.

They also wanted to understand how the herbal market operated. Most of the knowledge about herb teas rested with a few European herb wholesalers. These wholesale companies thought that nirvana had arrived—here were traditional American tea companies who knew absolutely nothing about herbs wanting to buy large quantities. The wholesalers looked forward to making a large profit off the Americans.

The tea-buying group at Lipton decided they wanted nothing to do with herbs. They handed over the responsibility to the research group. The research group wanted a professional taster who was comfortable working overseas. That was to be me.

Hal outlined my mission. I was to investigate and learn everything there was to know about herbs. I was to find out where they were grown, harvested, and processed. I was to have the freedom to travel anywhere in the world. I was to finalize my report within two years.

Not only was I being asked to travel the world again, this time I was being asked to apply my tasting skills and tea knowledge to the project. I had thought that RTOTI was the job of a lifetime, but it looked as if this one could be even better.

In the tea world, there are four important herbs: hibiscus, rosehips, chamomile, and peppermint; and one important spice, cinnamon. There are others of course, but those are the ones that interested Lipton. The tradition of using herbs goes back even further than tea. Women were the caregivers for the ancient tribes, where drinking teas of brewed matter was a normal, everyday occurrence. It was women who knew that a brewed concoction of the daisy chamomile made sleep possible and that the fragrant mint peppermint not only tasted and smelled good, it also settled stomachaches. This knowledge was passed from mother to daughter.

In China, the tradition of using herbs as stimulants, pacifiers, and medicinal aids is still in effect. I once fell ill on a visit to Yunnan and was taken to a Chinese doctor who looked into my eyes. Then he prodded, poked, and peered into various parts of my body. Through the translator, he told me that I was "out of bal-

Chapter 25

ance." I was given a long, tapered, slightly hairy root and told to set it in a glass of very hot water. I was to wait until the water cooled. I was then to drink the water, chew the root, and go to bed.

Few people experience the loneliness of sitting in a rural Chinese hotel, looking at a twisted piece of vegetable sitting in a glass of dubious origins, which was full of water with even more dubious origins, knowing you were about to drink and gnaw this stuff. But the affliction I had was similar to the onset of flu and every bit as miserable. I took the plunge and drank it down. The next day I was fine. My translator told me that my kidneys were out of balance and needed to be cleansed. I just know that it worked.

In Europe, the discovery of blood circulating around the body introduced the concept of science to medicine. That sent the Europeans down a different route to treating illness. As science took over medicine in Europe, the use of herbs declined. Only Germany remained an outpost of herb usage, and that country became the largest consumer, supplier, and processor of herbs in Europe. The knowledge of what grew where, how much it cost, and the seasonality of the herbs all rested within a few German companies.

In my new role at Lipton, I made Germany the first destination of my travels; it was the shortest route to knowledge. Once there it was made very clear that the German herb traders had no interest in sharing anything with me. Why should they? It was their competitive advantage. If Lipton wanted to learn about herbs, we were going to have to do it the hard way. I was going to have to go to the countries of origin.

I found that in the world of research and development, everything was about value. A research project had to have a payback. If there was no payback, there was no project. Lipton did not pursue science for the sake of science. They pursued science for the sake of commercial advantage.

Lipton was part of Unilever, the multi-billion dollar Anglo-Dutch conglomerate based in Holland and London, and making money was a golden truth. To justify the research in herbs, I had to show that the prices Lipton and Unilever were currently paying could be improved upon. Not just by a little, by a lot. There had to be payback.

I was based in the Lipton headquarters in Englewood Cliffs in New Jersey, and a special herb-tasting lab was built for me. It was also made clear that unless I could show that huge savings could be made, my lab and I would not last long. I reported to the director of beverage research. His four other direct reports each managed a research department, and they all held PhDs. We were an eclectic group, to put it mildly.

My task was to develop new, cost-effective sources for herbs that Lipton could use. The purchasing group within Lipton did not like my role. They saw getting new sources as their job, not mine. As far as they were concerned, I was a one-man band of trouble.

Hibiscus

I was summoned to Hal's office and given my first assignment. It was to be hibiscus, one of the most expensive and difficult herbs to obtain. It is obscure, growing mostly in remote parts of the world. Lipton used a lot of hibiscus bought from European suppliers, which was very expensive. It was felt that the European traders were making a lot of money on selling hibiscus to Lipton. I was to see if I could develop a direct source for the product. That was my mission.

Chapter 25

Hibiscus sabdariffa *in China. The production is very labor intensive.*

My first task was correctly identifying the herb itself. Hibiscus seems quite common—that pleasant bush that borders many gardens with large, droopy, long-stemmed flowers of radiant pink, white, or even purple. I soon found out that plant was the romantic "rose of China" (*Hibiscus rosa-sinensis*) and not the one used in herbal tea. That honor was reserved for *Hibiscus sabdariffa*. I read up and found that it was commercially grown in the Sudan, China, Thailand, and Mexico.

The flowers only last one day, which was okay, because it wasn't the flowers I needed. I wanted the calyx, the seedpod right under the flower. When the seedpod ripens, the plants are harvested by hand, then separated from the edible calyx.

This meant that lots of manual labor was required to harvest the calyxes, just like tea leaves. I learnt that after the first round of flowering, the wands could be cut to get an additional crop, but the calyxes from the second crop are much smaller and weaker in flavor than those from the first harvesting.

I knew that prices for hibiscus vary greatly from year to year. Like any agricultural crop, the prices depend on weather conditions, yields, quality, and demand. The most startling fact I learnt was that the total world crop was only about twenty-five thousand metric tons. I compared this to the tea output of Kenya, at three hundred thousand metric tons a year, and was stunned. Kenya alone produced twelve times as much tea in one year as the entire world's crop of hibiscus.

I reviewed the countries where hibiscus was grown to decide which looked like the best opportunity to develop.

In the Sudan, goat-herding nomadic tribes collected hibiscus. Sudanese hibiscus had a lot of extraneous matter, such as hair, feathers, stones, twigs, and dirt. The proximity to goats didn't help. The Sudanese hibiscus had a bright red color and a very acidic taste. It was extremely popular in Germany; almost 100 percent of the crop ended up there. That—and the fact that at the time I was looking at hibiscus, a nasty war was raging in Sudan—made me decided to bypass the Sudan. I'd had enough of guns and bullets.

My inquiries about China revealed that the hibiscus grown there was more sour than fruity in taste, but there was very little quality control among the farmers. All the samples I obtained were different, and there was no consistency in the appearance or taste. But China was eager to sell their produce on the world markets. My inquiries with the Chinese government resulted in a warm invitation to visit. I put China on the list.

My research about the Thai hibiscus industry revealed that it was a cash crop for the farmers. There was a traditional collection system in place. Middlemen, called collectors, visited the same

Chapter 25

farmers year after year and purchased the harvested calyx. They accumulated commercial quantities numbering in the hundreds of tons, which they resold overseas. The system worked well. Apparently the collectors had handshake agreements not to trespass on each other's territory. It seemed Thailand was living up to its reputation as being a very polite society, so I added it to my list.

But it was to Mexico that I turned for my first visit. I found out that virtually the entire hibiscus crop was consumed within the country. It was sold as extracts and in powdered drink mixes called Jamaica. I obtained samples and found the taste to be sweet, tangy, and delicious. My research discovered that the growing area was not far from the Lipton herb tea-packing plant in California. Mexico had to be my first choice.

I was intrigued with the logistics and the idea that we could purchase and process hibiscus in Mexico. Transportation and communication would be easy. It would save tens of thousands of dollars in storage and ocean freight costs, which we would have to pay if we brought in the product from China or Thailand. And because it was just down the coast from the Lipton plant in Santa Cruz, we would be in the same time zone. Keeping in touch would be much easier. It seemed a perfect match. I traveled to Mexico City and arrived on a rare day when you could see the surrounding snow-capped mountains. I took this as a good omen.

Mexico City is an amazing mix of European class and elegance on one side—wide boulevards, trees, gardens—and crushing poverty on the other. It is noisy and dirty, and the gas fumes are overpowering. It is also colorful, vibrant, and interesting. I made the mistake, however, of walking from my hotel to the address I had been given. It turned out to be a skyscraper overlooking the Rosa district of the city. I arrived with my eyes scratchy and sore from the fumes.

The men that met with me were polite. Their response to my idea that they could export hibiscus to the States had been fluent and fluid. "Come visit us," they wrote. "We need to talk." They wore starched white shirts and well-tailored suits. They were clearly part of the elite.

Amidst rich paneling and thick carpets, I was told that the peasants were not interested in selling or processing for Americans. This seemed odd to me; it was a terrific opportunity for the farmers. I tried to explain that this would be good for them. My remarks were met with polite smiles and exchanges in Spanish that I could not understand.

I was sent on my way with firm handshakes and wagged fingers. "No, no, this is not a business for you, *Señor*. Trust us, the farmers have no interest. Better you try the Sudan." Wagged fingers have never worked for me, and I decided to approach the farmers directly to see if we could do business one on one.

One advantage I had in my world travels was the wide reach of Unilever. Unilever made and marketed soaps, soups, detergents, and oils all over the world. No matter where I traveled, I knew there would be a Unilever sales office somewhere nearby. I made arrangements for a meeting with the farmers just outside of Guadalajara and asked to be picked up by the local Unilever agent.

My destination was a small village twenty-five kilometers outside of the city. Guadalajara was a bustling metropolis, but the

> **TEA TALES**
>
> There are two hundred different species of hibiscus, but only the roselle type, *Hibiscus sabdariffa*, is used for making herbal teas. Other types are put to a variety of uses including making paper and creating grass skirts, and some types are even used as offerings for the Hindu god Kali.

Chapter 25

countryside around was old Mexico—dusty streets, slinking cats, mangy dogs. Men in stiff white straw cowboy hats wore scuffed blue jeans.

I made my way there by taxi because my Unilever agent had not shown up. I had been told to expect a relaxed attitude about time in Mexico, but after waiting for two hours, I decided to take the cab.

I found the local cantina, and the farmers were all still there, but no Unilever agent. It was a dingy place, the door masked by a scruffy white sheet. The floor was packed dirt, and the only bright thing in the place was a neon sign for Cartagena beer. The farmers were gathered at one end of the room around tables and chairs. They had not taken off their hats and were empty handed. I offered to buy them beers; they shook their heads and waited for me to speak, their eyes serious.

My conversation with the farmers was painful because my Spanish is not even basic. I struggled through, suggesting that I could offer them a better price for their hibiscus, and I took out the samples I had brought with me.

Like farmers the world over, they loved to get the produce in their hands—to feel it, to crush it, to smell it. They handed the samples around, rolling the hibiscus calyx in their calloused palms. The samples got them animated, and a number of exchanges took place. I struggled to keep up, but it was impossible. My Spanish was not up to the task.

It was clear to me from their body language and the looks and turned shoulders that, as a group, they did not want to close the space between us.

Eventually one of the men stepped forward and handed me back my samples. The others fell silent as he spoke.

"It is too dangerous, *Señor*." His English was halting, but the finger he slid across his throat was a clear signal.

My mouth went dry.

His action caused a barrage of chatter, and I caught the word *muerte*, which I think meant death. It was at that moment that the light bulb went off in my head. I realized that what was good business for me and the farmers would not be good business for the polite, starched white shirts in Mexico City. The shirts probably owned the land that the farmers cultivated. They probably controlled the sale of the hibiscus or maybe even the distribution of the soft drinks that it went into.

I also realized why my Unilever associate had missed the appointment. He did not want any part of this interference with the natural order of things. I tried to ask questions about who owned the land, but I was met with stony silence and blank stares. Maybe they understood me, maybe they didn't, but it didn't matter; this was not going to work.

I packed away my samples and nodded to make sure they knew I understood that this business was not going anywhere. I smiled, shrugged my shoulders, and raised my palms up to the ceiling in the universal gesture of "oh well, I tried." It broke the ice, and the spokesperson pointed toward the bar. It was clear he was offering me a drink.

I grinned back at him, tapped my chest to let him know I was paying, and earned a wider grin in return. A few beers later I felt it was okay to leave, and I went out to my cab. The farmers all shook my hand but stayed inside the cantina. I understood. It was okay. If I had been in their place, I wouldn't have come out to wave goodbye either.

Chapter 25

I reluctantly shelved the idea of manufacturing in Mexico. Despite its convenient proximity to the Lipton plant in California, some things just were not worth messing with.

China, with its huge labor force and the desire to sell its agricultural produce into overseas markets, was the next logical option. Hibiscus was grown in the Amoy region of China, next to the Formosa Channel, which is now called the Taiwan Strait. To get there, I had to have special military permission, a guide, a translator, and a driver. When I arrived, I found that they were all security agents; they spent a lot of time checking on each other.

It would have been very funny if it were not all taken so seriously. They spoke broken English, but I am sure they understood everything I said. When I spoke to one of them, the other two would immediately move in to hear what had been said.

These were still the days of the Mao suit. This was supposed to create a sense of egalitarianism, but you could tell who were the more senior people. Their Mao suits were always made of better-quality material, with shaped waists and a nicer cut. My guides/guards/translators must have all been the same rank; the cuts of their suits and the materials were identical and looked mass-produced. This caused them great confusion because no one was in charge. It was like a three-ring circus.

The car was Japanese and had air conditioning, which I liked on full blast. My agents were not used to air conditioning; I noticed this when I heard a strange clicking sound and found it was the translator's teeth chattering. The three of them hated getting back into the car after each stop. There was a badly disguised tussle for the back seat, next to me and away from the vents. Eventually I took mercy on them and sat in the front and directed the vents at myself. Even so, at every stop they would leap out of the car to wave their arms and slap their bodies to warm up.

I was taken to a number of communes and shown huge acreages of hibiscus growing. The fields were a revelation: rows of single-stem hibiscus plants, with about ten blossoms per plant, stretched endlessly across the plain. Young women with scissors were collecting the blossoms. As I said earlier, the part used in herb tea is the calyx that supports the flower. The women cupped each calyx in their delicate hands, and then they snipped the stem with a curved iron blade. The worker then dropped the carefully preserved calyx into a wicker basket.

The problem was that there were no processing facilities. After picking the calyx, the peasants carried their haul in small baskets back to their homes. Their houses were wood and corrugated metal shacks with small gardens attached. The freshly picked hibiscus calyxes were spread on the dirt to dry in the sun. Happily frolicking among the drying calyx were chickens, the occasional piglet and its mom, and other domestic animals.

My attempts to explain that the material would eventually be considered a food product were met with shrugs. The translator raised his eyes to heaven and carefully explained in broken English: "It okay. We boil in water before drink." Pause. "Better you do too."

Cleanliness issues not withstanding, the price of the product made it very attractive. The challenge of communication, however, made it impractical. This was China in the late '70s. Trying to introduce good manufacturing practices into this environment would have been impossible and foolhardy. I reluctantly crossed China off my list.

At the end of my trip, in the last commune I visited, I was given a banquet. I had learnt by then that food is the most important element of life in China. Every roadside has beans, cabbages, or some form of green vegetable being cultivated. In China, there

Chapter 25

is not a square yard that has access to water that does not have something growing on it.

And when the Chinese eat something, they eat all of it. If it is duck, you get to taste the webbing between its feet; if they could eat the feathers, they would. And I bet the cooks would find some way of preparing the feathers so that they tasted great, for that is one of the joys of China. The food, whatever it is—and sometimes it's better not to ask—tastes fantastic. It tastes so good because everything is fresh. There was hardly any refrigeration in the countryside, so what was picked was cooked the same day.

The banquet was held in the main communal hall that had the ever-present picture of the chairman looming over us. Beneath the portrait sat a large, twelve-person round table with a lazy Susan in its center. This was spun around so that every dish was presented to you in turn. Plastic chopsticks were lined up on white porcelain plates in front of each person.

My visit was the reason for the banquet, but I was the least important person present; I was only the excuse for a food fest. My escorts all took seats, carefully positioning themselves so that they could look at each other. The head of the commune sat to my left, and I sat on his right, at the position of honor. To show his regard for me, he would lean forward, select a delicacy from the spinning centerpiece, and carefully place it on my plate. He would then smile, making a sucking sound to show how good this item was.

The workers in the kitchen must have been cooking all day. Dish upon dish upon dish, an endless supply of unfamiliar fare (after being told about the duck webbing, I had learnt never to ask). Young girls did the serving, and soon they had beads of sweat on their faces as they carried the steaming dishes back and forth from the clatter in the kitchen. I finally reached the point where I felt I could not continue eating, not even to be polite. My stomach was in distress. Belt loosened and top button undone, I felt the spasms of overindulgence flickering through my lower regions. When sliced oranges reached the table, I knew the meal had come to an end. Then, the leader of the commune leaned back, smiled at me, and gave an enormous, stomach-churning belch.

The belch resonated around the table, and all chatter ceased. The commune head was looking at me expectantly. I realized that I was expected to return the compliment. The moment dragged on; the silence grew. His look became quizzical, and I knew for the sake of Lipton and honor I had to do something.

My anguished stomach came to my rescue, and I let out an equally enormous sound—a fart. His eyes widened and then squeezed shut in a happy smile. I had enjoyed the meal, I was full, and that was all he asked of me. I had passed the test. That was not all I had passed, however, and the table quickly emptied as they sped me on my way, translators, driver, and guard in respectful tow.

After China, it was on to Thailand, continuing the search for a suitable source of hibiscus. The Unilever management in Thailand was entrepreneurial and eager to help the farmers. By this time, I knew how it was grown and knew the seasons and processes needed to get good dried product. I sat with two Thai Unilever executives in Bangkok and explained what was required. They took notes, nodded a lot, and said, "Yes, we can do this. Give us a little time."

The "little time" turned out to be three months, and I found myself back in Bangkok, inspecting a processing plant they had assembled. As Lipton could only buy ingredients that were cut into particles small enough to be packed into tea bags, processing the hibiscus was key to success.

Chapter 25

Machines like this have helped reduce the cost of hibiscus production.

They arranged for our first consignment of dried hibiscus calyx to be purchased from grateful farmers. We deposited the load into the hopper and held our breath as the first grinder clanked into life. The calyx ran out of the hopper into the grinder.

The stream of bright red particles plopped into a bucket conveyor, which carried the product to a series of sifter screens. The sifters started chugging, and small particles of hibiscus appeared—clean and free from dust. We waited until we had a pound and then brewed a batch.

It was superb, bright red with a tangy lemon note. We had done it. Lipton could now get hibiscus at a third of the price they were currently paying to the European dealers, yet the price included a profit to our processor partner, a cut to the Unilever department in Bangkok, and a very good profit to the farmer. It was one of the best win/win situations of my life. For the first time, the farmers began to see an alternative to the European buyer. I had improved their lot and done a great service to Lipton.

That success in the steamy suburbs of Bangkok proved the worth of what I was doing. I wrote a report to Hal. He circulated it among senior Lipton management. They were all pleased that their idea was working. I was given the go-ahead to proceed as fast as possible on the other four big herbs. I turned the hibiscus contract over to the purchasing department, where it was begrudgingly accepted, and turned my focus to rose hips.

Rose Hips

In doing my research for rose hips, I followed the same pattern I had established with the hibiscus project. I first read up on the origins and then sought out samples and trade information. I found that China, South America, and Eastern Europe were the main sources for rose hips.

The supply of rose hips was spread fairly liberally around the world, but China, a major grower, had been indiscriminate in using DDT. No one wanted to touch rose hips from China because of this, so I scratched China off the list.

Eastern Europe—a catchall phrase to indicate formerly Communist Europe—had plentiful supplies, but they also had long-standing relationships with the ubiquitous German companies. They were not responsive to direct inquiries on cost and quantities.

In those days, getting to Albania, Yugoslavia, East Germany, Poland, and Hungary with a U.S. passport was a problem. And once there, I'd be getting up the nose of the German buyers. Wandering the back roads, trying to source herbs, I would be making a general

Chapter 25

nuisance of myself. And no one likes a nuisance. Scratch Europe. That left South America.

I decided to concentrate on Argentina and Chile. Both had rose hip industries. Both were okay with American passports. I flew into Chile and stayed at a hotel overlooking the presidential palace. Salvador Allende, the president, had been deposed in a coup just a few months prior. You could still see the freshly covered bullet holes around the windows of the hotel overlooking the square. Military patrols were everywhere. It felt just like Uganda.

There were few tourists around, and the hotel management was clearly happy to see me. Most of my friends in the States were used to me going to odd and dangerous places, but even they looked sideways at me when I said I was going to Chile. They mentioned the recent revolution and the number of people who had disappeared. But a job was a job. I had no family to worry about; I figured if my time was up, it was up.

Chile possessed the ideal combination of plentiful wild rose bushes and a large peasant population. This made the collection and manufacture of rose hips a viable business. Chile was a smart place to start. There are many varieties of roses, but the rose hips found in most herbal teas are from the wild rose bushes grown in Chile, *Rosa mosqueta*.

The rose hip provides many benefits to mankind. It is one of nature's best sources of vitamin C. It makes a wonderful, gentle, sweet cup of herb tea. The small, shiny berries are pleasing to the eye. It can also be a joy for children because the small fibers and seeds on the inner shell of the rose hip make it one of the finest itching powders known to man. My sister and I used to make it in Kenya. We created the powder by stripping off the skin and collecting the seeds and the fine hairs. These we dried and then—in my case—dropped them down the back of my

These beautiful berries add not only color and flavor but lots of vitamin C.

sister's girlfriend's dresses. The seeds created an insane desire to scratch the places they touched. If I was successful, I was rewarded with screams and much twisting and contorting. I found this out when my sister's friends got their revenge and dropped the seeds down my back. The sensation went away after half an hour, but it was murder until then.

The rose hip begins life as a small, green berry that slowly ripens to a rich red-orange color by autumn. It is, ounce for ounce, much higher in vitamin C than oranges. Rose hips are common to all health food stores because they are also rich in vitamins A, B, E, and K.

In Chile, the wild rose bushes are scattered about the countryside. They do not grow in neat rows. These bushes are well over eight feet tall, with long, fierce thorns on the stems. The rose flowers are very small, but the rose hips are bloated and lus-

Chapter 25

cious. They hang in thick profusion, bending the wands toward the pickers.

The rose hips were harvested in a classic cottage-garden-industry style. Families scoured the countryside for the wild rose bushes. They gathered the berries by hand with crude rakes. They wore thick leather gloves to protect themselves from the barbs as they scraped the rose hips into tin coffee cans. Because the seeds never touch the ground, there was no concern about contamination from feathers, dirt, or any other extraneous matter as was the case in China. When a sufficient quantity had been harvested, the family took their bounty to a local collector. He paid them, and when he had accumulated a truckload, he sold it to a processor.

The industry was centered in southern Chile. Going there was a step back in time. Looking somnolent and weary from oppression, the towns seemed dazed to me. Thin dogs skulked alongside houses that huddled back from wide, dusty streets bordered by telegraph poles, from which drooped single lines of power cables.

In many ways it was like the area outside of Guadalajara, only poorer. In Mexico the farmers had worn high-stepping leather cowboy boots. In Chile they wore sandals. The hats were the same tough straw blend, but of a different style, rounder and with a wider brim. The eyes were the same, though—serious and attentive.

I was taken out to the countryside to watch the pickers at work. Like a tribe of gypsies, they arranged their tents and lean-tos in a circle. The remains of a large fire spoke to their nighttime entertainment. Very young toddlers stumbled about, watched by aged grandmothers. I did not see any old men.

In the distance you could see rows of bobbing heads making their way among the spiky rose bushes. The rest of the tribe were out in the fields, coffee cans in hand, raking off the ripe hips until the cans were full. The hips would then be dumped on a large canvas sheet, and the picker would go back to the bushes.

The air was heavy, the ground dusty, and the workers moved in slow, graceful circles, always making their way back to the canvas sheet. The pile of hips grew steadily until they were spilling over the edges of the sheet. Then a halt was called to the work. The men pulled the corners of the sheet together, fastened it through the eyeholes, and then secured the sides. I guessed there must have been two hundred pounds of rose hips in that bundle. A second sheet was pulled out and spread on the ground, and the process began all over again. It was like something out of the Middle Ages.

Any industry dependent upon peasant labor is at risk once wages and opportunities start to increase. Such is the case now in Chile. The towns are no longer sleepy backwaters, and the rosehip industry is in peril. No peasants, no crop.

Harvesting is in rhythm with the seasons. Picking begins in early March and is over by the end of April. Once the rose hips reach the processing plant, they are mechanically pre-cleaned. This eliminates leaves, branches, and other debris before the hips are dried.

The old-fashioned way of drying was to spread the rose hips out on a hard, flat surface in the sun. This method was very cost effective; the sun burns brightly and often in southern Chile. However, to standardize product quality as well as speed up the process, drying chambers and tunnels are now used. The fresh fruit is placed on trays, where it is dried by gusts of hot air produced by fans and heat exchangers.

Chapter 25

> **Tea Tales**
>
> Rose hips are simply the fruits of the rose plant that form after the flowers are gone. They are extremely high in vitamin C, with around two thousand milligrams per one hundred grams of dried rose hips. They are frequently used in a variety of folk medicines and are even the basis of a popular soup in Sweden called *Nyponsoppa*.

Next the rose hip is separated into three distinct product categories. Whole dried fruit is used in aromatic adornments and potpourri. The cracked shells are exported to Europe and used in loose blends of coarse-cut herbs, fruits, and flavors. They are also used in instant soups and juices as well as in pharmaceutical extracts.

Ground shells are destined for the tea bag trade. Their very dense nature is of great use to blenders. A conventional twenty-foot container of black tea—broken orange pekoe, fannings grade—will weigh about ten metric tons, whereas a twenty-foot container of rose hips will weigh eighteen metric tons, almost twice as heavy. The hips are packed in bulk sacks or paper bags then stuffed into containers and shipped down to Valparaíso, the famous Chilean port, for export around the world.

There was no need to try and work directly with the growers. The industry was already established. All I needed to do was negotiate a good price with the processor and ship a trial consignment to Lipton.

But I did not have purchasing authority; I only had research authority. With the support of Hal, I got the purchasing manager to buy a consignment and ship it to the States. It was a straight shot from Valparaíso up the Pacific Coast to San Francisco, and from there it was only two hours to the Lipton plant in Santa Cruz.

I ran a trial costing. The company would save hundreds of thousands of dollars annually. I wrote another report. Hal grinned. I smiled. Even the purchasing department joined in the applause. After all, they had placed the order.

I wondered how much of the purchase price made it back to the peasant families I had seen filling their coffee cans.

Two down, two to go. Or so I thought.

As Lipton is owned by Unilever, Hal had passed on the reports on rosehips and hibiscus to them. I suddenly received an invitation to coordinate my efforts with a Unilever buying group. It meant that I was no longer working alone; I was probably going to have a Unilever employee tagging along.

I was very lucky in the companion they chose. Henri de Venevelles, a French nobleman whose forefathers had lost the family fortune playing cards, worked for Lipton France, another Unilever Company. He was also a member of the task force, and he wanted to investigate rose hips out of Argentina. As Argentina was also a major world supplier of chamomile, another one of the herbs I was supposed to source, Henri and I agreed to meet in Buenos Aires. He turned out to be an affable man with a great sense of humor and a relaxed view of life. I could not have asked for a better traveling companion.

We met with the local Unilever personnel, and we were told about a place called El Bolsón, high up in the Andes, that was overrun with rosehips. But they warned us that it was very remote and well protected. We did not know what that meant, but it did not dissuade us.

We flew across the Argentina Pampas to Baroloches, a ski resort area for the wealthy from Buenos Aires. In Baroloches, which had

Chapter 25

The strange and extraordinary El Bolsón valley.

a heavy German influence, we rented a car for the trip. We had no need for air conditioning, as the air was crisp and cool. We had no need for a radio, as we were too high in the mountains for any reception. So Henri and I chatted as we drove on for hours on dirt roads into the high Andes until we finally reached El Bolsón.

El Bolsón was a shock: a microclimate buried in the midst of the mountains. It consisted of steep valleys dotted with Swiss chalets with bright red window boxes filled with geraniums and edged with carved wood. It could have been Bavaria. Turned out it was.

At the end of the World War II, thousands of Nazis who did not want to meet the Allied forces obtained Argentinean passports and disappeared. A lot of them ended up in the most remote valley they could find. Henri and I had arrived among them.

We met our guide—a six-foot-three, blond, blue-eyed giant who bowed slightly and clicked his heels when he met us. His name was Juan Ramirez, and he was about twenty-five years old. I guess he was the offspring of one of the refugees who took shelter there in 1945.

We toured the valley, and it was teeming with rose hips—more than enough to ensure a year-round supply—but the valley lacked a labor force to harvest the hips. It also lacked any sort of processing facility.

The only business in the valley that had transportation arranged to the outside world was a trout farm. It was run by a stocky German in his late sixties who took us around the fish ponds and the processing plant. He had a very small autoclave unit that he used to pack his fish. The canned fish were collected twice a month by a truck that made the trip from Baroloches. It was a small truck, and he did not think there would be room for any rose hips in it. Beyond that, he did not talk much and seemed happy to send us on our way.

That evening, eating German sausages and drinking beer in our Alpine chalet, we told Juan that we did not think it would work. The valley was too remote.

"*Ja*, that is good," he said.

"No, that is bad," we said.

"Bad for you but good for us," Juan replied.

Henri and I, of mutual accord, took a mouthful of sausage and chose not to answer. The beer was excellent, which did not surprise us, nor did the oompa band music that was piped out from the behind the bar.

The valley truly was a beautiful place—lush, green, with steep mountains on either side that soared into the heavens. It was overcast during our visit, but beams of sun shone through cracks in the clouds and created a sense of serenity and peace. We saw very few people. I believe they stayed indoors while we were around. We

Chapter 25

saw no schools, no school children, no teenagers. Juan had been very reticent about life in the valley. He had spoken a lot about soccer, and I had learnt they had two soccer teams that played each other. We were told a lot of retired people lived in the valley and that it was a very quiet place.

We left the next morning for Baroloches. As we wound our way back down the mountains, Henri and I shared our thoughts. Neither of us had felt in any personal danger on our visit, but both of us felt tense and uneasy; we were thankful to have made it out of Gestapo valley.

Chamomile

Once back in the civilized world of Buenos Aires, we enjoyed a huge steak dinner and told our Argentinean Unilever associates of the strange visit to El Bolsón. They acknowledged that they had heard rumors about the place, nothing more. We took cold comfort in their assurances that they would have come looking for us had we not returned.

They had arranged for us to leave the next day for Patagonia, where the chamomile crop was grown. We faced a seven-hour drive through the Pampas flatlands, so we left early. One of the local men was to accompany us to act as a driver and as our guide. Contact had been made with a few farmers who knew we were on our way.

Chamomile is one of the most widely consumed herb teas in the world. An attractive daisy with startlingly white petals and a bright golden center, chamomile is used in Europe as a medicinal tea for its calming properties. It also has a wonderful character; superior chamomile has a taste strongly reminiscent of honey.

We knew that the European traders had set up processing plants in the growing areas of Argentina, and we hoped to visit

Chamomile has been used for thousands of years as a soothing herbal remedy.

them. When we got there, we were told that it was not convenient to visit, and we were on our own. It did not surprise us. We had expected difficulty in getting into the factory to see the machinery.

What we wanted to learn about was the sifting process. Each composite flower consists of many small whole flowers or florets. Each floret has all the parts of a full-sized flower. It has the petals, the stamens (the male organ), and/or a pistil (the female portion). The pistil, when fertilized, produces a seed. All in all, it's a pretty sexy package, you might say.

However, it all needs to be separated and sifted. This requires either air separation, which is achieved by using the weight and density differences between the petals and florets, or screens. We wanted to know which was used.

Chapter 25

> **Tea Tales**
>
> *Matricaria chamomilla* is the type of chamomile used for making herbal tea. It is one of the oldest herbal remedies still in use, going back well over two thousand years. In most herbal teas, it is used for its calming effect, but it is frequently used to settle upset stomachs, as well as for its antiseptic properties.

Chamomile is an adaptable plant and will grow virtually anywhere. It is an annual and must be planted every year, but it typically self-seeds. This makes it the ideal crop for the smaller farmers. We knew there were other farmers in the area who grew chamomile. They might tell us what the processes were.

We drove to the town nearest the plant and found a coffee shop. From there our associate made his telephone calls to other farmers. In the meantime, we showed the waitress our samples of cut chamomile and asked if she knew anyone who grew this. You never know when you might get lucky.

Before we could finish our second cup of espresso, a tall, lean, well-weathered man entered the shop and approached us. He introduced himself as Herr Vogel. We invited him to take a seat as we handed him our samples.

Herr Vogel was one of the local Unilever contacts, and he grew chamomile. He knew exactly what we were looking for. He had his own processing operation and said that he would be very pleased to show it to us. Henri and I had hit pay dirt. We happily followed Mr. Vogel out to his farm.

His facility was a large shed stuffed full of baled chamomile with a cutter, grinder, and sifter placed in the doorway. This was necessary because the air was filled with chamomile particles. The workers wore face masks, and it was just as well because you could hardly breathe next to the machines. Henri and I lurked near the door. In the few minutes we had been in the shed, we had become covered with a fine dust. It looked like a very unhealthy place to work.

We saw that a series of screens separated the whole heads from the stalks. The stalks were diverted into a miller and cut up to tea bag size. Not a portion was wasted.

The florets and the petals then entered a separate stream until they too were split into two drums. We knew the percentage of stalk to florets determined the price for the raw material—high stalk content meant a cheaper price. This was because different levels of volatile oils produce varying degrees of aroma intensity and depth of color.

Just like tea, there can be many blends of chamomile that are all designed to suit the buyers' needs.

All chamomiles on the store shelf are *not* the same. If you want to know how good the chamomile that you are buying is, take a teabag, tear it open, and look for shreds of petal. If you can see an abundance of white petals, you are enjoying a good-quality chamomile. If it is a mass of short sticks and stems, then it is the cheap stuff.

We asked Mr. Vogel how he blended the final product. He said he waited for the buyer to tell him what level of quality was needed. He took our sample and analyzed it for us. Apparently we had a low-medium blend—about 70 percent stalk, 15 percent cut petals, and 15 percent florets.

He volunteered to improve it for us. We asked him how much that would cost, and he told us. The savings over the current price we were paying were in excess of 300 percent.

Henri and I agreed that this was a good idea and then started to talk about density. The density of chamomile has been the bane of most tea bag packing plants. Chamomile is very fluffy. Putting

Chapter 25

fluffy product into small tea bags has caused many a production manager anguish. It refuses to be packed, or when it is packed, it causes a tea bag to resemble a plump pillow. This problem is solved by blending proportions of petals, florets, cut stalk—and even, on occasion, seed—to achieve the right density.

By this time it was growing dark, and we suggested adjourning for the night and resuming the next day. Mr. Vogel insisted on taking us back to his home for a drink. We followed his truck to a magnificent hacienda-type lodge that was approached by a long driveway. An airstrip ran off into the distance. Henri and I agreed that he was doing all right with the chamomile business.

Over drinks we discovered that he also had a cattle business and a slaughterhouse. Herr Vogel was doing better than all right; he was doing well. We also learnt that others in the business used air separation for their chamomile, but this created problems with dust blockages in the suction pipes. He explained that was why he used the old-fashioned sifting system.

The next day we left our associate to drive the rental back to Buenos Aires while Henri and I, courtesy of Herr Vogel, flew back in his private plane. We had made a new friend, and a trial shipment of chamomile was easy to arrange. It arrived in San Francisco in good condition, it tasted great, and it packed like a dream. The operations people at the plant were happy. The purchasing department was more than happy. Another report hit Hal's desk, and I hit the road in pursuit of cinnamon.

Cinnamon

The sweetness of cinnamon makes it an important ingredient for herb teas. The fact that cinnamon pricing was greatly influenced by a New York spot market made it a target for our project.

Spot markets can be good for you and then can hurt you. Essentially they represent a stockpile of goods that is readily available to a market. The goods are sitting in a warehouse and can be picked up immediately. This is important if the product is only grown in a tropical climate and the transit time to the United States is six weeks or more.

Spot markets are very convenient, but they can also be very expensive. If there is a sudden demand for cinnamon, the price will rise. If the demand slackens, the price goes down.

> ### Tea Tales
>
> The cinnamon spice is actually the bark of a small evergreen tea native to Sri Lanka and southern India. Its flavor is due to an aromatic essential oil that is very strong despite making up less than 1 percent of the bark. A Dutch captain once claimed that the scent of cinnamon growing on the shores of Indonesia was still detectable almost twenty-five miles out to sea!

This price volatility made cinnamon a difficult ingredient for an herb tea blend because you never knew what the price was going to be. If you had the resources to bring in cinnamon from overseas and warehouse it yourself, then you could control your costs and your quality.

It did not take me long to uncover that all commercial cinnamons are the same genus but different species. *Cinnamomum zeylanica* is the true cinnamon from Sri Lanka and is generally preferred by Europeans and Latin Americans.

Learning more about Ceylon cinnamon was easy. I had spent a lot of time in Ceylon and knew the country well. I discovered that the true cinnamon was also truly expensive and in short supply at origin. An aggressive foray into that market would only make it

Chapter 25

A worker bundling cinnamon in Indonesia.

more expensive. The Ceylon cinnamon also had a very distinctive taste, sharper and sweeter than that from other origins. I thought it would make a unique blend, but the supply issues made it much too risky to use.

Most of the cinnamon in America came from China or Indonesia. The first was *C. cassia* and the latter *C. burmannii*. The *burmannii* is called Korintji because it is grown in the Korintji region on the west coast of Sumatra in Indonesia.

Cinnamon comes from an evergreen tree that can live up to one thousand years. Not only that, but you can start harvesting the cinnamon bark four years from the day it is placed in the ground as a seed. Those are the kind of yields farmers love.

One of the great attributes of cinnamon is that it is dense. Many herbal ingredients are light, fluffy, and difficult to pack—chamomile, for example. Not so with cinnamon. Once the bark is cut to tea bag particle size, it has the same flow characteristics as black tea. This makes it an excellent component for herb blends because it is also sweet and aromatic.

I thought Indonesia was the place to start with cinnamon. Indonesia was a major supplier of tea, and I had good contacts there. I arranged for samples to be sent and planned a trip there to investigate further. I flew to Jakarta, on the island of Java. There I made contact with my Lipton associates and was flown to the big island of Sumatra to pursue cinnamon. We drove out to the village where the cinnamon trees were being harvested. Sumatra was hot and humid, and the heavy green vegetation hung motionless with not a whisper of a breeze.

We approached the village through rice paddies on either side of the dirt road. Neat squares of grass bordered rice plantings meandering along a slight valley. Thick jungle edged the rice paddies. The village was on a slight rise, with stilted huts no more than three feet off the ground squatted on either side of the street.

The farmers wore their traditional sarongs—long, wide, brightly colored strips of linen—wrapped around their waists. They went bare-chested and wore sandals. I envied them. In my long khaki pants and a short-sleeved shirt, I tried to stand erect so that the sweat would run straight down my legs and not encourage my trousers to stick to my body. Every movement brought forth a droplet of moisture from my body. It was beyond hot. It was Hades. The car was blessed with air conditioning, but as soon as we exited the vehicle, any sense of comfort was lost. The farmers, my ever-ready Unilever associate, and I were drenched in sweat.

The villagers were excited to see us. They came out of their thatched-roof shelters that were open on three sides to catch whatever breeze found its way through the thick foliage. The women were topless as well. Having been raised in Africa, this

Chapter 25

was a common sight to me and no big deal. Not so to my Unilever associate, who had not had the benefit of an African upbringing. The farmers were amused by his interest, as were the women.

The farmers demonstrated how they coppiced, or topped off, so the tree would sprout shoots. They then showed me how the collector marks the bark at one-meter lengths and then cuts a circle around the base of the shoot and at the tip of the meter mark.

These wands were then carried back to the huts where the bark was removed. The collector sat on the floor, a pile of wands by his side. Leaning forward, he effortlessly slipped a wand alongside his big toe. Taking a curved blade, he made a notch in the bark and with one fluid motion sliced a two-inch wide sliver of bark from the wand. Placing the sliver off to his side, he rotated the wand and repeated the exercise. Within a minute, the wand was naked and replaced by another. I could only admire the athleticism of the cutter. Bent like a gymnast, he worked his way through the wands, which were then laid in the sun and dried. As the wands dried, they curled to form the familiar quill shape we associate with cinnamon. We left the village with much handshaking and smiles. They had enjoyed our visit. I had enjoyed it as well, almost as much as I enjoyed the air-conditioned ride out of the jungle.

There are different grades of bark depending upon how far up the tree you harvest. The grade depends upon the amount of oil the bark contains. Cinnamon's volatile oils are very light, and the aromatics evaporate when in contact with air. The volatile oil content is important to the buyer because of the cost—the higher the percentage of oil, the higher the asking price.

My challenge was to process the bark at origin in Sumatra and capture the oils in the product before it lost its flavor. To process at origin required a machine that would crack and then cut the hard bark into small pieces. Another machine would then sift the pieces into sizes I could use. The machine would have to run at a slow speed, as a high speed would generate heat and "burn" the cinnamon. This had never been tried at origin, as far as I knew.

First I had to find the machines and a mechanic to run them. Then we had to establish specifications on size and oil content and a means of testing the specs we had established. It all took time and money.

To run a worthwhile trial, I needed at least one ton of cinnamon. The farmers were thrilled at the prospect of a cash sale. I purchased the cinnamon and had it shipped to Jakarta. I could not find any milling equipment in Sumatra; I am sure it was there, but by this time my body was melting. We ran the cutting trial in Jakarta, and I carried the samples back with me to headquarters in New Jersey. When we arrived, the blenders rejected the cinnamon as burnt. I decided that cinnamon might be one of those herbs where the effort was not worth the reward. My plan, which would have provided Lipton with a significant savings, proved to be a failure.

Not much processing of cinnamon has been done at origin. It was feared that cutting the cinnamon at origin would lose volatile oils, and this would reduce the flavor of the cinnamon. Not only that, I had confirmed through my trial test run that, if you were not very careful, you could easily burn the cinnamon when cutting. In the end I decided it was safer to rely upon the spot market in New York for cinnamon rather than try to circumvent the system.

The incentive for the farmer to process cinnamon and make it more valuable was also less intense. All of his production was being purchased. He had trees that lasted one thousand years. It was also so hot and humid; why would anyone want to work that hard? I had a great deal of sympathy for this viewpoint. The memory of that heat remains with me to this day.

Chapter 25

I wrote a report for Hal outlining my thoughts that we should stick with the spot market and not change the way we were doing things. I got it back with one phrase scrawled in red pen across the top of the page: "You can't win them all."

Peppermint

Legend has it that Persephone was not a woman to mess with. When she discovered that her underworld husband, Hades, had a thing going with the beautiful nymph Minthe, jealousy burned within her. She worked her powers and changed Minthe into a lowly plant. Hades couldn't undo Persephone's spell, but after consulting the oracles, he softened it a little so that the more Minthe was trodden upon, the sweeter her smell would become.

Over the millennia since Minthe was shortchanged, her name has changed to *Mentha*, and the word became the genus name for mint. Few people consider it a lowly plant. *Mentha* is a staple in many businesses—toothpaste and chewing gum, for example, as well as herb teas.

However, if someone tries to sell you a packet of peppermint seeds, grab onto your wallet and head for the door. Unlike the sexy chamomile, peppermint doesn't produce seed. It is a sterile hybrid of *Mentha spicata* and *Mentha aquatica* and is mainly propagated by cutting and dividing. These are the only sure ways of propagation that make certain you get the plant you want—and in the business of mint production, getting what you want is critical.

As an herb tea, peppermint has some wonderful attributes. It has been used for hundreds of years as a beverage. It is freely available in Europe and America; supply has never been an issue. It also has the distinct advantage of being a single herb that requires no additional flavoring. It can be packed all by itself, or it can be line-priced with other herb blends that cost a great deal more. Peppermint is the most profitable of all herb teas.

Like tea, mints have a distinctive taste depending upon where they were grown. The Balkan and Adriatic countries are major suppliers of peppermint. Most of the mint produced in the United States comes from the Pacific Northwest (Washington, Oregon, and Idaho). Mint cultivated in the Midwest (Indiana, Michigan, and Wisconsin) tastes different from that grown in the Pacific Northwest. The mint grown in each of these regions has its own characteristics, caused by the local climate conditions, soil types, and so on.

Peppermint is a perennial crop that grows back each year from its stolons (underground stems). The crop is harvested just before it flowers so that it is full of oil. Oil content is what makes the plant valuable. The mint season begins in the Pacific Northwest in June and continues through August or September, depending on the weather conditions. After it is cut, the peppermint is allowed to sun dry. It is then collected by combine harvesters that deposit the cut peppermint directly into trucks for transportation to the oil extraction plant.

For herb tea, the processing involves cutting, sifting, and sorting in order to remove the stalk and stems from the raw leaf. My issue with peppermint was not the supply but the taste. Each growing country has a different taste; getting the right one proved to be a challenge.

A traditional solution is to blend a mix of origins. I knew that was a difficult task because that's what I did as a tea taster and blender. It took a lot of experience and knowledge of origins to be able to match the exact taste year after year. The better answer was to find a single source that produced consistent quality.

Chapter 25

> **TEA TALES**
>
> Peppermint is often regarded as the oldest herbal medicine, with archaeological evidence showing its use around ten thousand years ago. It is also widely accepted as an effective ant repellent. Not only will the ants leave, but your house will smell nice, too!

At the time, the United States had an extremely sophisticated and very large peppermint industry. It was based entirely on producing essential oils for the cosmetic and food industries. Through a friend who had once tried to cultivate tea up in Oregon, I was introduced to a group of peppermint farmers who also had an extraction business.

The idea of using mint leaf for tea was new to them, but they liked the idea. Farmers are an independent, philosophical, humorous, tenacious, and money-minded lot. How can you not be when your livelihood depends upon hard work and the whims of the weather? They are also as tough as nails and as hard as old wooden planks; they take a lot of convincing.

I flew up to Portland and headed east. I crossed the magnificent mountains of the Cascade Range and came down over the foothills of Mount Hood. I followed Highway 26 across the dry plains over to the Warm Springs Indian Reservation.

The farmers had chosen this place to meet, as it was a place of distinguished beauty that they thought I would enjoy. The land and the rocks reeked of ancient tales. You could just feel the history lurking in every turn and twist of the road. The raw, simplistic grandeur of the place made you feel small. It had a rugged beauty; there was nothing manicured or cultivated. It was nature at its best.

Perched high among the craggy rock outcrops of high-plains foothills, the Paiute tribe had built a casino. It was there that I sat across the table from a group of peppermint farmers and tried to work out a deal.

The farmers were a mixed lot. Affable Joe, Taciturn Dick, Hard Mark, Brilliant Doug, Friendly Bob: I anointed them with nicknames in my head to keep track of them. As I still deal with them to this day, the names have stuck, and they still apply. They were summing me up as well, in their quiet way, and I must have passed muster because we moved onto serious business discussions after a few pleasantries.

The peppermint farmers quickly picked up on the idea that we could expand their product line beyond oil. They agreed to do test runs and needed about three months to assemble the equipment. I agreed, and we shook hands on the deal. I returned to New Jersey to wait for the call.

Almost three months to the day, it came. I flew back to Portland, and because I had enjoyed the drive up Highway 26 so much, I did it again.

I was directed to a barn hidden away outside the small town of Madras. They pulled the wide barn doors open, and I was met with the sight of an amazing machine they had assembled. Over thirty feet tall, it resembled a pyramid with a series of cascading sifting screens protruding out at odd angles. These caught the peppermint leaves as they cascaded down through the screens, which sorted them out and moved them on. At the base of the pyramid, large collection tubs received all the different leaf sizes. It was an amazing contraption.

The machine was driven by an eight-feet-wide, sixteen-feet-long stapled rubber belt that drove a gear assembly. This jarred and vibrated the screens and created an immense noise. I watched as

Chapter 25

the machine shook and rattled the leaf down through the system; the peppermint dust rose in the air in great clouds.

But the machine worked. It may have looked like a mad scientist's dream machine, but the result was a spectacular leaf style. Good peppermint should taste slightly hot and slightly spicy—the peppermint from Washington and Oregon had exactly the right impact. If we could get the costs where we needed, it was a go.

We ran the numbers. It was clear we could save a lot of money buying directly from the farmers. "Could the machine keep up with production needs?" I asked.

"This machine is over forty years old. It may be rickety, but it really works. It is so simple that if it breaks down, it is easy to fix," was the response from Affable Joe.

I placed a trial order on behalf of the purchasing departments, and we were in the mint business.

As you can imagine, a tea taster with an English accent did not rate high on the scale of manly professions for these farmers, but in the coffee shop meeting following the successful trial run, I noticed one of my remarks had caught the attention of Hard Mark. Mark was a stocky, tough negotiator who had led a lot of the discussions. When I mentioned that I often carried an iron when I went on the road, he glanced in my direction.

"You travel with an iron?" he asked, resting massive forearms on the coffee counter.

"Wouldn't leave home without it," I declared.

"What type?"

"Well, it has a folding plastic handle," I said, demonstrating with my hands how the handle pulled up and over.

"Smart," he said. "Do you declare it?"

"No, I just pack it," I replied.

"I pack too," he said. I could tell he was warming up to me. He leaned forward, smiling. "I've got a Ruger Security Six .357 Magnum out in the truck. Never leave home without it." He grinned. "What caliber is your baby?"

I hesitated. I had put a lot of work into this peppermint project. I could see that it was all about to blow up in my face.

"It's a traveling iron. I use it to press my trousers," I said quietly. There was a long moment of silence.

"Press your trousers?" he asked.

"Razor sharp," I said.

"Razor sharp, eh?" he managed to get out.

"Could cut you," I added.

"You could be dangerous," he said before dissolving into a fit of coughing and laughter. "You could be trouble." He started to pound the table with the palm of his hand. When he finished choking, he explained the joke to the others at the table, and I endured another five minutes of hilarity. For better or worse, I had a nickname—Iron Man—that has lasted to this day. It has quite a ring to it, as long as you do not know the origins of the tale.

I wrote another report to Hal. My mission had been accomplished. Hal informed me that management had decided to use a mix of the herbal sources. They wanted to use both the European suppliers and the sources that I had developed. It was a compromise that kept both the Europeans and the Lipton purchasing group happy. I was to be rewarded with a promotion to keep me happy as well. I was sent back to the tea-buying group as manager of tea buying and blend development.

Chapter 25

Hard Mark and Iron Man comparing irons.

Chapter 26
The Secret of Blending

BLENDING TEA IS A LOT LIKE COOKING FROM A RECIPE—FIRST YOU HAVE TO HAVE ALL THE INGREDIENTS YOU NEED. FIFTY PERCENT OF A REALLY GOOD TEA AND FIFTY PERCENT OF A REALLY BAD TEA WON'T WORK—YOU END UP WITH BAD TEA. TRY THE SAME MIX WITH MEDIOCRE TEA, AND YOU HAVE THE SAME PROBLEM. A POOR TEA WILL BRING THE GOOD TEA DOWN TO THE LEVEL OF ITS BLEND MATE.

There is a lot of rubbish tea around the world. When tea is made, it is split into grades; the best are called primary grades. The teas that fall between the cracks are called off-grades.

You can tell an off-grade by tearing open a tea bag and looking at the tea. An off-grade will have lots of small pieces of wood, stalk, tea fiber, and dust in it. Most of the time, the tea will also be a dirty brown.

Some tea-producing countries actually outlaw the sale of tea fiber. But in other countries, it is sold as tea. It is cheap and it legally comes from the tea bush, and once the tea is iced and sugar is added to it, the awful taste is thinly disguised.

In the world of tea, tasters are not necessarily blenders, and blenders may not be buyers. I have had the opportunity to

Chapter 26

experience all three disciplines. Of the three, blending is the most challenging.

My days at Brash Brothers had taught me that I enjoyed the challenge of tea blending. I described tea blending as baking a cake with ingredients that are ever-changing. At Brash Brothers, the challenge was constant but on a small scale. At Lipton, the scale was huge.

If you do not have all the ingredients you need, you must substitute. But if you do use a different tea, you have to adjust the taste in endless combinations of components until you get it right and it meets the standard. If you enjoy this mental and tasting challenge, it is a lot of fun. If you don't, it is hellishly frustrating.

I learnt early on in my new role that the Lipton blend was defined by the purchasing activity six months before the tea ever arrived in America. The correct mix of teas had to be bought. Some teas have substitutes that can be used easily. The North Indian teas that dominated blending fifty years ago have now been replaced by Kenyan tea. High-grown Ceylons can be replaced by Indonesians or South Indian teas. There will be many who will be offended by these remarks, but the reality of blending is reflected in the import figures into the United States. Half a century ago, tea was mainly from India and Ceylon; now it is Argentine, Indonesian, and the like. Things change. Such is life.

Some of these changes are man-made, and some are not. Argentina, which now provides America with almost 40 percent of the tea it consumes, has no tea pluckers. Machines harvest all of the tea produced in Argentina. This makes Argentine tea cheap. It also makes it very efficient in producing the same quality of tea year after year. Only weather conditions and the state of Argentinean finances can affect the supply of tea from that country.

The Secret of Blending

Taxes drove many tea buyers out of India. India has struggled to recover the U.S. market.

This is a man-made change that is helped along by the type of tea that Argentina produces. Argentine teas have a deep reddish color and their own distinctive character that does well in iced tea. And 80 percent of the tea drunk in America is iced tea.

Other changes can be brought about by political ineptitude or arrogance. In the early '80s, Lipton used to import a fair amount of Indian tea. Toby had knowledge of India and its teas; he liked the country and its people, so we purchased Indian tea.

Then the government of India introduced an export duty and said it applied to all tea exports. At the time, Lipton was sitting on a lot of tea awaiting shipment. We had bought it at auction, and it was worth a lot of money. Toby was told that he had to pay duty on this tea before it could be shipped.

"Surely not," Toby said. "We purchased this tea long before the law was passed."

Chapter 26

> **TEA TALES**
>
> The higher the altitude at which tea is grown, the shorter and slower the growing season. This makes the flavor much more intense, which is why places like Darjeeling in India and the Yunnan province in China are both valued for their flavorful teas.

"Not so," said the Indian government to Toby, who flew out to discuss the matter with the authorities. "The duty is retroactive and applies to all teas awaiting shipment."

"Not fair," said Toby.

"Who cares?" the Indian government said. "Our country, our rules."

Guess what? Lipton started to buy Kenyan teas (better, in my opinion, than the Indians), and India never got its share of the American trade back. It was gone, and a look at the U.S. import figures today shows that it is still gone.

In the '70s, mainland China teas were unknown in the United States. Then, in the '80s, they took off and became dominant. In the '90s, they sank back again as the business ethics of the free-wheeling Chinese traders became clearer to the buyers.

The Chinese traders became known for antics such as ignoring a contracted price if the market moved up. This is frowned upon in the tea trade. A contract is an honored document. To refuse to deliver a contract simply because the price you committed to now looks low compared to the current market is simply not done. It is a way of ensuring that buyers will go elsewhere for their needs the next time around. And that's what happened.

Then Vietnam, which in the '90s became a dominant force in the coffee business by planting millions of coffee bushes, suddenly promised to do the same for tea. These days we see Vietnamese teas appearing on the lists of imports.

The Secret of Blending

Many of these teas are used for iced tea. Color is very important in constructing an iced tea blend, and it must be carefully maintained. Taste is important, of course, but color rules the day, closely followed by price.

Hot teas are almost the opposite. They are sought for their flavor; color is a secondary need. The traditional tea-growing countries such as Ceylon and India all cultivate and sell excellent teas, but they are not always suited to iced tea use, so their market in America is limited. The high percentage of tea solids from these origins, when extracted in the brewing, precipitates into a murky cloud when the tea chills. It does not look attractive.

So as the consumers in America shifted to drinking more iced tea, the origins shifted as well. My task as the manager of blending and tea development was to oversee this shift and make sure we always had the right mix of teas on hand.

I did not mind being back in the corporate world. To tell the truth, my traveling boots were becoming worn out. I was ready for some time in the home office. And at Lipton, there was a constant stream of visitors from overseas to enliven things. This was understandable because we were the largest tea company in America. All of the tea planters who ever harvested a bush wanted to sell us their teas, and we were unfailingly polite in having lunch with them. Gossip was still the lifeblood of the industry, and news from overseas was always welcome.

Offer samples of teas from around the world came in every morning and were brewed up. My first task of the day was to taste these teas and see if they met our needs. If a sample met our quality and price limits, I would make a note and put in a bid. Sometimes I could do this by simply picking up the phone and making a local call if the sample came from a local tea broker. If the sample came from overseas, then I had to send a fax and await a response. If the

Chapter 26

bid was successful, I had to enter the purchase into the inventory mix and see how the balance came out.

My days were spent balancing the incoming inventory from over twenty origins. There were three tea bag packing plants and one instant tea plant, and making sure they all received the right amount of tea at the right time was a challenge for the three tea buyers whose job it was to manage it day-to-day.

I enjoyed my days. Blending always challenged me. I was still traveling overseas to the tea-growing regions, only this time I was traveling as a tea buyer.

The power of this position is hard to imagine. We had the responsibility of spending millions of dollars, and we had only each other to judge the wisdom of our purchases. Yet we always looked out for the best buy, for the best quality. As a buyer you are at the pinnacle of your profession when you work in a role like that for such a large company.

Chapter 27
Taking an Iced Brew to Market

Having settled into the new role, I began to look at blend development. Lipton had always marketed a good tea that could be brewed both hot and iced. We sold bags for restaurant use designed just for iced tea. But for home use, the public got a blend that worked both ways—it could be brewed up hot and also prepared as an iced tea.

Eighty percent of the tea consumed in America is drunk as iced tea. Once tea is chilled, the flavor and character become much more difficult to taste. This has allowed a lot of really cheap, awful tea to be offered as "iced tea blends." You can find these blends on the supermarket shelves and in mass merchandisers—one hundred tea bags for ninety-nine cents. That is under one cent per bag! Do you think that the tea bag contains good tea? Nope, it is your basic rubbish.

The nature of iced tea has allowed a multitude of sins to go unchecked. I once saw my wife, who is from Texas, boil up a pan

Chapter 27

of water, throw in two family-sized tea bags, and watch them boil away happily.

"What are you doing?" I cried.

"Making iced tea," she said.

"That's not the way you make iced tea," I pleaded. "You will boil it to death."

My wife paused, looked at me with one of those looks that only a husband who has transgressed knows, and uttered the words that ended the debate right there. "It's the way my mama taught me."

"Ahh," I muttered as I retreated to fight another day.

A really good tea made into iced tea is a pure delight. It is crisp, refreshing, and energizing. Lipton was challenged by the cheap tea revolution. Their yellow-box blend was designed as a hot tea blend that also made really good iced tea. We called it an AC/DC blend, capable of serving the needs of both the hot tea and iced tea drinker.

I was asked to develop a blend that was specifically designed for iced tea use.

Iced tea drinkers love color. The color of their glass of iced tea has to be "right"—it cannot be too dark or too yellow. It must have a reddish tinge to it. It cannot be too light in color; that signifies weakness.

However, the perception of color has to match the perception of strength, and strength can come in many forms. Astringency is strength, pungency is strength, aroma is strength—but get the wrong balance, and the blend will be too much of one thing and not enough of another.

I also faced the challenge of the large Lipton volume. To produce a tea for the largest tea packer in North America meant one had to have a year-round supply of components. Most importantly, the tea always had to match the standard. Once in the marketplace,

Taking an Iced Brew to Market

Tasting teas at Lipton. Compare this scene to the photo in chapter 6.

the tea always had to taste the same. There could be no deviation in quality.

A lot of superior teas do not make good iced tea—they cloud. Solids precipitate as they chill down, and the glass turns cloudy. This is not what you want to see in your glass of iced tea. "Hard" water will also cause cloudiness in a tea, though brewing with bottled spring water or distilled water will avoid that problem.

A combination of clear-brewing yet strong teas was needed from year-round sources. Some of the strongest teas, with the most color, are grown in the Assam Valley. They make hopeless iced tea, as do high-grown Ceylons and Darjeelings, because of the clouding issue. The very benefits that make these origins create good tea—their high percentage of tea matter in solution—makes them poor candidates for iced tea blends. Getting the right mix of origins was the challenge.

Chapter 27

During the twelve months I was working on the development of the blend, I learnt how to use a focus group. A consumer focus group is a tool that has become essential for developing new products. A target group is selected that meets the demographics of the group that management wants to capture. Our groups were invited into rooms with one-way mirrors. I sat behind the mirror with others and listened to people discussing the tea, how it tasted, and what they wanted.

After a blend is created, a test market is selected. This is normally a region of the country where the product category is popular. For example, it would not make sense to test market an iced tea product in Maine in the winter; Florida in the summer is a much better choice.

During the test, sales are carefully audited. Certain magic thresholds must be crossed before the product is allowed into the mass market. Then it is slowly rolled out across the nation. This was the case with our blend.

Where did we find the tea that went into our iced tea blend? This I cannot tell. Even today, that must be kept secret and confidential—as per the non-disclosure agreement I signed when I joined Lipton. I developed the Lipton iced tea blend that is still sold in America today; I was thrilled that I had the chance to complete the task for them, but what it consists of must remain forever secret.

The iced tea blend was a success. I was secure in my profession, I was having fun, and I traveled a lot. My career at Lipton was rolling in the right direction. I wasn't really happy living in New Jersey, but I left a lot on business trips. The situation wasn't so cheerful for my wife. I had married a Texan who hated the Northeast with a passion.

Taking an Iced Brew to Market

Then, as if in answer to her prayers, I received a phone call from Barney Feinblum, president-in-waiting for Celestial Seasonings. He asked if I wanted to join in the leveraged buyout of Celestial. He and four friends were going to buy the company for fifty-five million dollars. He thought I would be a good mix in the management team because I knew herbs, I knew black tea, and I knew corporate infighting. Was I interested in moving to Boulder, Colorado, to join Celestial as an owner and a part of the management team? Of course I was.

But Lipton was interested in keeping me. On a bright Saturday morning, I drove in to have a private meeting with the senior vice president of my division. I was told that Lipton did not make counteroffers, but they were going to accelerate my career path by ten years. I was told the terms and conditions, and my head spun. I called my wife on the way home and gave her the news. There was a long silence, and then she said, "Honey, I don't know what you are going to do, but the kids and I are going to Colorado."

With that, I said goodbye to Lipton for the third and final time.

Chapter 28
The Celestial Years

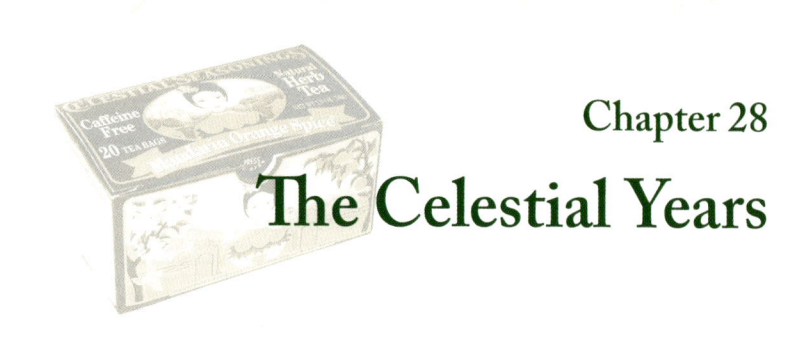

T*HE RIDE OF CELESTIAL SEASONINGS HAS BEEN WEIRD, WIRED, AND EXTRAORDINARY. IT IS THE STUFF OF LEGEND. CREATED BY MO SIEGEL AND JOHN AND WYCK HAY, CELESTIAL STARTED A BEVERAGE REVOLUTION IN AMERICA. TEA AND TEA PACKAGING HAVE NEVER BEEN THE SAME SINCE THE 1970S WHEN CELESTIAL BURST ONTO THE SCENE.*

The early days were hippie heaven. Stories abound of female production managers who would show their breasts to crews if they would work an extra hour of overtime. It was the '70s, and Boulder was rocking. Work was fun at Celestial, and they had a product that just hit the spot and created a new beverage category—flavored herbals that tasted good and were reputed to do you good as well.

Kraft, the food giant, took notice and purchased the company in the early '80s. This began an uneasy marriage—mainstream corporate America meets tie-dye. It did not work. What did work was Kraft's advertising. It drove Celestial from ten million dollars to fifty-five million dollars in sales per year.

Kraft tried to make the marriage work. When executives visited the Celestial plant in Boulder, they were told to blend in. "Ditch the suits, wear blue jeans." But when the Kraft executives appeared in newly bought blue jeans that creaked, topped off by black wing-tipped shoes, it was obvious that something wasn't quite right.

When it was clear the herb tea phenomenon was real but the marriage between Kraft and Celestial wasn't, Lipton stepped in and tried to buy Celestial. The deal was hung up in the courts by Bigelow Tea Company. Bigelow claimed the sale would give Lipton a monopoly on the herb tea business. The dispute was tied up in court for over a year. At that point, Kraft decided to sell to a management buyout team, and Lipton dropped out of the picture.

I watched the attempted acquisition by Lipton with great interest. When the call came from Barney asking me to join the leveraged buyout group (LBO) as an owner-manager, I said yes. I was to be vice president of research and development. I arrived with my family in Boulder on Thanksgiving Day of 1988. It was cold, wet, windy, and gray, but nothing could dampen my mood.

That changed when I found what state the company was in. Corporations live and die on gossip. The Kraft sale to Lipton had been in the works for a year. The court case had been going on for an additional six months. For eighteen months, no work had been done at Celestial. People had been standing around in the corridors and gossiping about their fate.

There were no new products in the pipeline or on the drawing board and no new systems. Half the management team had gone back to Kraft, and the money had left with Kraft as well. We were an LBO, and it was a shock to my system. Lipton had been warm, fuzzy, and secure; Celestial was highly insecure. And as an owner, my life savings were on the line.

Barney demanded I jumpstart research and development and get some new products out into the market. New products are the lifeblood of companies; we needed the excitement and cash flow

Chapter 28

they would generate. The one new category I felt was going to take off was organic. The organic category fit Celestial's image of being natural and real. I decided to develop a line of organic black teas.

There were no organic teas in the market prior to Celestial. Many of the herbs that Celestial used were wild-crafted. This meant they were picked in the wild—in many cases making them more organic than organic—but wild-crafted herbs did not have a piece of paper certifying them as organic.

In an agricultural world that demands fertilizers and pesticides to increase yields and profits, organic products are a halfway house. The misunderstanding is that "organic" means the product was grown without the use of fertilizers or other additives. This is an incorrect assumption. "Organic" simply means that the use of certain organic, natural fertilizers is approved and kept track of and that a documentation trail is carefully maintained of everything that is applied to the product. There is an association that charges to come out and verify the documentation trail. The product is then given a seal showing that it has met the organic criteria.

To my mind, wild-crafting is truly organic. Fruits and herbs that grow wild are harvested in the wild. No pesticides or fertilizers are applied, but there is also no record of that. And fruits and herbs grown in the wild can be subjected to the wind-borne drift of pesticides and other chemicals. Wild-crafted products cannot be claimed to be completely free of chemicals, only that they are free of applied chemicals. Celestial's wild-crafted herbs could not be used for any organic blends.

Price is also a consideration for any organic production. Tea is no different. If you grow organic tea, the yield of tea per bush goes down. In order for an estate to stay level in cash return, a higher price must be paid for the organic tea.

When an estate converts to being organic, the taste of the tea does not immediately change. Buyers are asked to pay a higher price for a tea that tastes the same as one grown with chemicals. In order for buyers to pay a premium on organic tea, they must be able to charge consumers a premium as well. Is organic tea worth the premium? In my opinion, yes, but not necessarily for the subtly improved taste, for that takes years to develop. I believe organic production is worthwhile for the good of the soil, birds, bees, workers, and nature in general. But in the late '80s, organic teas had not yet hit the marketplace.

> **LOCAL LORE**
>
> The Priyar Tiger Preserve in Kerala is one of 27 tiger preserves in India. It is home to 62 known species of mammals—many of them endangered—and while the tigers are their most famous inhabitants the park is also the only sanctuary where Asiatic elephants can be seen in the wild.

I thought organic teas would play well to the Celestial consumers. They liked natural products, and in 1989 organic was a new concept for most mainstream companies. I thought the consumers would appreciate the subtle difference and pay the premium that organic teas would require. Launching an organic blend for Celestial would also make us first in the marketplace. The marketing team at Celestial agreed with me, and the sales team thought they could sell the idea, so I was not alone in the decision. I wanted to keep the idea away from my competitors, so I began my search for an exclusive, private, almost secret source of tea.

The place where I found my organic tea was in the southwest corner of India, near Kerala. The estate is in the midst of a pristine Bengal tiger preserve, an incredible slice of land that defies

Chapter 28

Tigers are amazingly well camouflaged, and you are very lucky if you see one.

description. It is devoid of people; that is a remarkable thing in India, where it is nearly impossible to get away from people as they are everywhere at all hours of the day. India teems with people, but in the tiger preserve, peace reigned in the ancient forest. Soaring trees draped with vines covered the jungle floor with dappled light. Years of fallen foliage had created a soft carpet of leaves so you moved silently through the forest with barely a shuffle. Birds screeched above as they warned others of your presence. It was a magical place.

The estate was reached via a long, winding dirt road, which opened up into a series of rolling hilltops on which sat the tea bushes. The estate had stopped using fertilizers four years prior. It was depending entirely upon mulching to sustain the tea bushes. Large pits had been dug along the roadside and were full of grass cuttings, weeds, and anything else that would rot. The heavy smell of decay permeated the estate. I did not find it unpleasant because I knew the purpose of the smell.

The first time I tasted the tea, I thought I had a winner. A clear amber liquid swirled into my spoon, and the fresh aroma reached me before I raised the spoon to my lips. Fresh, distinct, full of flavor, and bursting with aroma, the tea was good, really good. The infusion, the wet leaves of the just-made cup of tea, was a bright coppery color, a sure indication of high quality and good manufacture.

I had brought along some standard teas to taste the organic tea against. It was not a contest. Next to the organic offering, the other teas looked and tasted dull and plain.

Of course the organic tea was not cheap. Not only that, it had to be handled and transported in such a way that it never came into contact with anything nasty. You can imagine the uproar if our pallets of precious organic tea were shipped to the United States in a mixed cargo of chemical dyes or lead paints. Special shipping was needed, and that was an extra expense, but it was all worth it in my view.

In 1989 Celestial organic tea was launched nationwide. It sold well at first. Like so many of Celestial's ideas, it was before its time. We did not have the advertising dollars to tell the public what the product really was, and it languished on the shelf.

Barney then asked me to develop a really good line of black teas. I told the board that if a company famed for its herb teas decided to produce a line of regular teas, those had better be the best teas ever. On the crowded tea shelves, a consumer has multiple choices of black teas. If the customer selected a Celestial black tea, it would be to try it out. We wanted our customers to come back again and again, and we could only achieve that by being the very best tea on the shelf. The board told me to do it.

Chapter 28

I did, and in the early '90s, Celestial Seasonings' English Breakfast blend was voted the best of its type in the United States by *Consumer Reports* magazine. The blend was sensational. We were the first herb tea company to be so honored. I think we remain the only one.

What makes a tea blend sensational? It is like eating a tomato grown in your own garden, one that is picked when it is full, plump, and oozing with juice, so that when you reach out to pluck it, all it needs is a gentle twist before it settles into your hand, warm from the sunshine. Compare that to one of those tomatoes grown in California that are picked when they are green and then gassed for preservation and shipped around the country. They have as much flavor as a piece of cardboard. That's the difference between freshly picked tea and tea that has been sitting around in a warehouse for a year or so.

The key is to use really fresh tea—blend it immediately and send it into distribution within days of blending and packing. Most tea companies take a minimum of eight months to get the tea from the bush to the cup; we were able to do it in three.

Next you must use really, really good tea—all tea is not created equal. The problem for the consumer is that in dry leaf form, all tea looks the same. All tea grown in Argentina, for instance, is mechanically harvested. This results in a lot of stalk and fiber. All Argentine tea factories have very sophisticated sorting and cleaning systems, so the black leaf that emerges is very neat and clean. This does not mean, however, that it can compare in taste to a high-grown Ceylon or a Kenyan tea in cup quality.

The essential to producing a really good tea is to monitor quality continually. Be rabid about every single purchase, and make sure each one matches the standard you have established. We were mad about quality control. So many companies establish a standard blend and then just buy against the standard instead of looking at every single batch of tea that goes into the blend.

When these three things come together, you get a really good tea—and we did. We had a really good tea. But you can no longer buy that blend. It was changed after I left Celestial; I could never figure out why. I know that it was not cheap to make, but, as my grandmother used to say, you cannot make a silk purse out of a sow's ear.

Another recent method of growing tea and herbs in a carefully controlled environment—and one that I would have liked to pursue for Celestial had it been available at the time—is biodynamic farming. Biodynamic tea is an even more extraordinary development in the world of natural products. It resonates with the quirky world of Celestial.

Biodynamic farming is a way of growing food in which no artificial fertilizers or chemical sprays are used. All influences on the plant and the soil are considered—things such as light, warmth, and the water in the soil and atmosphere. Even the sun, moon, planets, and stars are all brought into the growing equation.

This concept of growing food was developed in the 1920s by Austrian Rudolf Steiner, a man who wore many hats—philosopher, literary scholar, architect, playwright, and social thinker. He described a method of making a series of biological preparations that would influence soil development.

The biodynamic methods are practiced in Europe, the United States, and Australia by devoted followers. In some boardrooms, biodynamic farmers are considered the lunatic fringe; in others, they are thought to be the path to the future. Biodynamic farming takes us into the ancient past of cultivating in tune with the waxing moon and the rising tide and using rhythms and methods that

Chapter 28

Packing fertilizer for biodynamic tea production.

are both hokey and yet so advanced no one can quite explain why they work. But work they do.

Australia is the largest producer of biodynamic produce, with over two million acres of land being cultivated in this fashion. The biodynamic tea estate I visited was a happy place, fully living up to its creator's view that the farm is a single entity and what sadness is felt there will permeate into the ground as much as will joy and laughter. Biodynamic farms are not required to be communes, but I think the spirit that is evoked in such farms creates a sense of unity and commune. Certainly the tea estate I visited had such an atmosphere.

The roads leading to up to the Idulgashinna tea estate were raked and clear of rocks; the water runs by the roadside, clear of weeds. Close-cropped grass was underfoot everywhere we stepped, and a feeling of well-being was evident. Each of the buildings and functions were clearly marked with black on white wooden signs: kindergarten, maternity ward (for the workers), horn shed, worm shed (for the creation of fertilizer). All were connected by flower-bordered pathways.

In biodynamic production, teas are planted by the cycles of the moon and harvested to the rhythm of the earth. Fertilizer comes from cow and worm droppings.

The worm farm is on-site. Every day, deep concrete tubs full of worms are churned with rich organic soil matter to keep the occupants happy and active. If you are a worm, this is definitely the place to be.

To harvest the droppings, two pounds of worm-ridden matter are scooped up in mesh and suspended over a deep bucket. Over the course of a week, moisture seeps through the matter and drips in a steady stream into the bucket. It is mixed with cow manure that has been "matured," and this combination is packed into hollow cow horns and buried in shallow pits for five months.

The horns are then dug up, their contents mixed with water, and the bushes sprayed with the mixture. It was how they sprinkled the bushes that caught my attention. Tree branches are cut and trimmed, and then workers push their way through the tea bushes, dipping the tree limbs into diluted worm droppings and waving them above their heads. A fine spray of refined, diluted worm poop then settles gently on the tea leaves. The liquid is a potent concentrate of goodness for tea roots—the bushes love the stuff.

All the weeding is done by hand with long rakes, and insects are controlled by inter-planting flowers that produce scents to drive them away. This holistic approach to tea growing works in dramatic harmony with nature and produces extraordinary results; these are some seriously good teas.

Chapter 28

Sounds weird, I know, but tea bushes treated this way shimmer with health. The teas taste delicious. Biodynamic teas are purchased in Europe by consumers who are prepared to pay the much higher premiums they command. They are even more expensive than organically grown tea. Biodynamic teas have yet to make it past the supermarket buyers' desks in the United States; I am sure the price scares them off. It's a shame because the American consumer is missing out on something special.

Chapter 29
Celestial in the Rearview Mirror

My career at Celestial was successful. I had developed a line of black teas that had seen success in the marketplace—one of which had been voted the best English Breakfast tea in North America. That was quite a coup for an herb tea company.

I had introduced the first organic black tea blend into national distribution in North America. With the assistance of Dr. Mary Mulry, a very accomplished director of research, we had assembled a great team of personnel. They, in turn, created a number of herb tea blends that entered the top ten sellers for the company. Entering the top ten sellers is the gold standard for success. All was well with my world. I was secure in my role of tasting and blending.

Then Barney offered me the role of vice president of operations. He did this as the operations group was floundering in inefficiency. Despite the efforts of the management team to work their way out of the problem, it was not going away. Barney decided he had to change the head of operations staff, and he offered me the role.

Chapter 29

I accepted because I wanted Barney's job, and the best way of getting that was to make a success of operations. Such is ambition.

I trimmed the operations group by eliminating a number of positions that were not needed, and I shifted the right people into the right jobs. We all then created a vision and focused the team on manufacturing efficiently. It worked—things were going well.

That was not the case overall for the company. Laden with debt, Celestial was struggling. After three years of strife, the largest investor invited Mo Siegel, the original founder of the company, back into the fold. As part of the deal, Mo was to invest a considerable sum of money. As Mo was running an organic cleaning company at the time, Celestial acquired that as part of the deal for him to return as well.

There was jubilation at the company. Mo was seen as the savior, and in many ways he was. His investment helped out with the debt restructuring, and his personality and reputation gave Celestial a much-needed boost.

Mo started to reorganize his management team. He wanted me to remain in the company, but not as head of operations. For that role, he had selected a manufacturing professional, John Hine, who had formerly been at Frito-Lay. John explained that both he and Mo wanted me to stay in the company, but in a purchasing role reporting to the head of operations.

The only job I was interested in at Celestial was Mo's role. As that clearly was not available, there was only one place for me to go, and that was out the door. I was on my own again, but I had a plan.

Life and luck had positioned me perfectly for what I had in mind. Lipton had exposed me to just about every source of herbs in the world. Celestial had taught me how to blend and develop herb teas. Under the brilliant and mercurial Barney Feinblum I learnt the discipline of running a multi-faceted company. I knew how to process and cut the herbs. I had the knowledge, and I was ready to strike out alone again.

This time, however, there were other elements at play, and I could not be alone in making this decision. In the course of my life, I had married twice. The first marriage, to an English girl I had met in Uganda, lasted two years. We were great friends but should never have married. Penny was petite, luscious, dark-eyed, gregarious, and funny. In a land of few white women, she was a rare prize. We had both been raised in Africa and knew the customs and culture. Away from Africa, it was different. We fit each other as friends but not in other roles. Our parting was sad but necessary, and we remain friends to this day.

My second marriage was to Sandy, a girl I met in Texas. Sandy was tall and slim and ravishing, with raven-colored hair and eyes that were blue, gray, or green depending upon the day and her mood. Part of her family had come to America in the 1600s and fought in Bacon's Rebellion. She has the blood of the Cherokee Nation on both sides of her family. We met in the early 1970s and eventually married in the mid-1980s. The result was a family of five and all the trappings that go with life in America—numerous cars, credit cards, and lots of debt.

We lived in an old rambling house set alongside the county line. There was nothing but green fields and open space between us and the mountains. We had a pond, three acres, and a mortgage to make you blush.

I had a plan, but I needed the support of my wife to execute it. The risk was failure, which would mean bankruptcy and the

Chapter 29

loss of the house, our cars, and all the trappings. I needed to make sure such an event would not involve the loss of my wife and kids as well.

The only way for my plan to work was to cash in all my savings to start a business. The considerable amount of Celestial stock I owned was useless; it was highly restricted and could not even be used as collateral. If I was to succeed, it was going to be the old-fashioned way—by taking a risk.

My wife did not hesitate; she said yes. But we had three children—ages seven, four, and one—and we needed a backup plan in case of failure. We decided that if the venture failed, I could always mow lawns in Texas—Texans love English gardeners.

We had a Ford LTD station wagon we called the "Bitch-Mobile," so named because Sandy had once cut off a couple of women in a dash for a parking space, and when she returned, she found "bitch" scratched on the car door. The survival plan was to pile the kids in the car and head for Texas; it remains our backup plan to this day, although the Bitch-Mobile is long gone.

I was ready to take the risk. I had a willing partner in crime. All that was left was to execute the plan.

Chapter 30
Sandbar Sets Sail

MY PLAN WAS TO PROCESS HERBS AND SPICES AT ORIGIN AND SELL THEM DIRECTLY TO TEA COMPANIES. IT WAS THE SCHEME I HAD DEVELOPED FOR LIPTON THAT THEY HAD HALF-ACCEPTED.

Lipton had decided to split their herb business between the traditional wholesalers in Germany and the new sources that I had developed. In my mind I felt they had half-accepted the concept, but I could offer the same concept to tea companies around the world.

In the small world of tea, people knew who I was and what I had been doing. When I approached major tea and herb packers with my proposal to sell them better herbs for less money, they all agreed this was a good idea. They wanted to see samples and pricing.

Sandy and I formed the Sandbar Trading Company, and I set off to the Far East. The old Lipton hibiscus operation in Thailand had closed down. The owner was now making rubber gloves.

"Big market. Lots of AIDS around. People want rubber gloves." It was a succinct business plan, and there was no persuading the operator to go back to hibiscus, but he would sell me the equipment.

Chapter 30

I bought everything I could lay my hands on. Then, with a Thai trader I knew, I set up another processing plant. The equipment took months to assemble, but when it was finally in place, we ran hibiscus, lemongrass, and ginger through the machines. I then hand-carried the samples back to the States and Europe to sell the idea.

The samples tasted great, and the buyers loved the prices. I collected the signed contracts as I went from country to country. I then flew back to Thailand to set up the processing plan. We were in business!

Sandy covered home base while I covered the rest of the world. I traveled the cheapest way possible, buying bucket seats out of London, always flying in the back of the plane and always, it seemed, in the early hours of the morning.

The wolf was at our door, and failure was unthinkable. I had a six-month severance package from Celestial because, as I was an owner of the company, even a resignation was treated respectfully. So I had to make a success of the venture within a year. Otherwise I would be cutting grass.

Contracts in place did not mean cash flow. It just meant that whenever we delivered the product, we would get paid. As these contracts were from six to twelve months in the future, immediate cash needs had to be met.

On my way out to Thailand, I stopped in Ceylon to make contact again with my old tea friends. I was asked what I was up to and always gave the same reply: "I run the Sandbar Trading Company." Occasionally I would get questions as to whether we imported anything other than tea. The first rule of owning your own company is that you can always do what is asked of you, so I said, "of course." When the requests came in—be they for tea, herbs, or frozen food products—I faxed them to Sandy.

"I thought we were in the herb tea business," she said.

"We are in the trading business," I said. "Can you get a quote?"

"Oh, sure," my raven-haired beauty said without a moment's hesitation. "I know a trader. I'll give her a call."

When I presented the quote, I included a very nice markup for Sandbar Trading. Two days later, I had signed contracts from Stassen Foods for large quantities of frozen chicken parts.

The trouble was that I didn't have a clue how I was going to pay for them. Payment could only be collected on delivery of the products, and no food company was going to extend credit to a new business. I approached a bank to do a finance deal, but they were of similar mind about extending credit to a new business with no assets.

> ## TEA TALES
>
> It was while watching a cup of espresso being made that I had the idea for the tea pod. After a lot of experimentation, I identified certain leaf sizes that produce a fantastic cup of tea using the same application of heat, high pressure, and water. The key is that the important components of tea—flavor, caffeine, phenols, and color—are rapidly soluble under high pressure when wet; however, some of the undesirable aspects of tea, such as bitterness, take longer to go into solution. This means you can make a fantastic cup of tea (even iced tea) in forty seconds. I thought the idea good enough to patent it (U.S. Pat. No. 5895672), and I now own the worldwide rights to the technology.

So Sandy and I cashed in the 401K—our final reserves—and paid for the consignment. Six weeks later, I awoke in a cold sweat. I had dreamt that the freezer unit on the containers had failed and the food was melting. But it was only a dream; the chicken

Chapter 30

arrived in good condition, and we received payment along with more orders. We were alive and well in the trading world.

Stassen Foods was part of a much larger organization that included banking, tea estates, milk production, battery manufacturing, and liquor. It was not long before the natural connection between my herb business and the Stassen tea business came to mind.

Together we decided that there were great opportunities if we combined our talents and our sources, and Stassen North America came into being. This presented Stassen with a source of herbs and me with over thirty-five tea estates and a tea-packing factory. We were soon heavily involved in developing and packing teas for many supermarkets, coffee chains, and retail outlets. We do so to this day.

From there to my own brand was an infinitely more difficult step, but it was one that was in keeping with the evolution of the business. If you are not growing, you are dying. There is no such thing as standing still.

We never stood still.

Chapter 31
Putting a Face to Cooper Tea

THESE DAYS I HAVE A SECOND COMPANY, THE COOPER TEA COMPANY. ALL OF OUR PRODUCTS CARRY MY NAME—AND MY FACE. IF YOU DIDN'T KNOW BETTER, YOU MIGHT THINK THAT'S AN ENTREPRENEUR'S EGO RUN AMOK. YOU'D BE WRONG.

All strong brands seem to have a good story behind them. That's one of the things I picked up while slaving away at Lipton and Celestial Seasonings. The products are good products, but just as important is the brand and what it means to the public. Successful products have brands that excite consumers, attract a following, and eventually turn people into loyal customers.

Often this fan base is developed through word-of-mouth stories about a unique event in the company's history—stories based in truth. Lipton Tea and Celestial Seasonings both had good stories. Thomas J. Lipton owned tea estates in Ceylon, but while he imported tea into the United States for years, he did not become famous until he lost five challenges for the America's Cup. His yachts, all named *Shamrock*, lost, and he became a beloved

Chapter 31

figure in America for both his attempts to win and his grace when he constantly lost. So I guess you can become a big-time winner by being a big-time loser! Celestial Seasonings has the folk tale of Mo and Wyck and their wives picking herbs in the Rocky Mountains, packing them in hand-sewn bags, and driving around the Midwest hawking products out of a VW.

Behind these folk tales are other elements of finance, distribution, product development, market research, and all the other trappings of a business. However, the essential romance of the story remains true and reflects the personality of the brands and the people behind them. The public responds to that personality, and suddenly the product becomes authentic and real and thus seems more valuable.

It is interesting that in the cases of both Lipton and Celestial Seasonings, individuals are closely associated with the companies. Twinings is another famous brand identified by an individual. Move beyond tea, and you find Paul Newman with his salad dressings, Orville Redenbacher with popcorn, Colonel Sanders with Kentucky Fried Chicken, and Frank Perdue with Perdue Chicken. Lending your image to a product definitely gives it that extra dose of credibility. Customers know that you don't put your name or face on a product unless you believe in it. And in my case, it's certainly the truth. These days I have products that are available all across the nation, and they all have my face and my name on them one way or another. I am proud of the really good teas that I have created, and I want people to know that I stand behind them. The flagship product is B. W. Cooper's Iced Brew Tea. And, like Lipton and Celestial, B. W. Cooper's has a pretty good story behind it. Unsurprisingly given the events in my past, it's a story about how luck can play an amazing role in one's life.

Putting a Face to Cooper Tea

My African heritage gets a big nod via the bush jacket, safari hat, and thorn trees.

It was 1999, and I was in Dallas meeting with the beverage folks at the 7-Eleven Corporation. They told me that sales of the tea sold on the fountain dispensers alongside Coke, Pepsi, and the other soft drinks were not doing well. Consumers were not buying the tea. The 7-Eleven management thought it was the taste of the tea that was turning the consumers away. They asked if I could develop a good-tasting iced tea that could be served through the fountain dispenser. Oh, and it needed to come in a standard 2.5-gallon box of concentrate that yielded 30 gallons of great-tasting tea. I wasn't yet sure what all that meant, but I said yes. I didn't have a clue about fountain dispensing, but I did have a clue about how to make good tea. I figured if I knew the latter, I could learn the former.

To me product development is an equation. There are a number of parts to the problem. Each part relates to the other, and they

Chapter 31

all relate to a single answer. In this case, the answer was to be a delicious all-natural tea concentrate. I just needed to figure out how to get there.

The first thing I learned is that fountain dispensers use boxes of concentrates. These concentrates are mixed with water at the fountain. So a 5:1 concentrate means that five parts of water are mixed with one part of tea to produce a beverage. Thus a 2.5-gallon bag of 5:1 concentrate will produce 15 gallons of finished product (2.5 x 6 = 15). Since most people put enough ice in a fountain cup to fill up 50 percent, that yields about 30 gallons of tea to be sold.

With the science worked out, I then turned to the art involved—making the tea itself. Not much has changed inside my heart when it comes to blending tea. I still get a thrill walking into a tea room. I still am excited by the challenge to create a new blend or a new concept, and this challenge represented both. No tea man could ask for a better problem to solve.

I first looked at what was being sold in the market. I discovered that the two leading national brands had between them eighteen separate ingredients. These "drinks"—I could not bring myself to call them teas—included tea made from powders, artificial flavors, artificial colors, and all kinds of additives. I also knew from my tea-trading background that the type of tea generally purchased for tea concentrates was the least expensive tea you could find. So the answer to that part of the equation seemed pretty simple. No matter what you do to cheap tea, it will still taste cheap. To get a premium tea concentrate, start by using premium tea. I use very good teas.

Several questions still had to be answered, though, in order to reach the final answer. How strong do you make a tea so that it still tastes great when diluted in a cup 50 percent full of ice? What is the optimal process needed to produce an all-natural tea? How do you pack this all-natural, unique tea concentrate so that it has a practical shelf life, both in storage and after being opened? And how much would it all cost? I could produce the best tea concentrate in the world, but if it cost ten dollars a glass, we weren't going to have many customers.

None of these questions had easy answers, and it took me much longer than expected to work the equation through to completion. But at the end of four years, I returned to Dallas with a sample of the world's first all-natural tea concentrate, and it had a shelf life of nine months unopened and a quality life of six weeks once opened. I couldn't wait to share it.

The 7-Eleven management tasted the tea, and grins immediately spread across their faces. "Great-tasting tea," "This is real tea," and "Like my Momma makes," were some of the comments I registered. It was a hit! Then came the question I wasn't expecting. "What are you going to call it?" I had been so focused on developing the product that it never occurred to me that a brand name would be needed to sell it.

"7-Eleven tea," I stuttered.

"We don't want a 7-Eleven-branded tea. This is a great tea—you need a brand," was the quick response. And my world started spinning because I knew better than anyone how much it cost to develop a brand. It was one of those moments that defined my life. If I said yes, I was committed to a course that I knew was going to be expensive, full of risk, and with no guarantee of success. If I said no, I was safe. I was only out four years of work and product development. Wisely I hadn't yet bet the farm. "OK," I said, and promptly bet the farm.

"We like the idea that a real tea expert developed this tea. You could play around with that concept," were the final words I heard as I left the meeting. After a few homegrown attempts at

Chapter 31

branding, I got smart and hired a marketing firm in Boulder. The Sterling-Rice Group was small, elite, and very good. They visited 7-Eleven with me and worked via web conferences where graphics and concepts were discussed. We finally whittled the concept and artwork down to six logos and designs. We flew down to Dallas and presented our combined work. A decision on the logo was reached. The concept of "Iced Brew Tea" was settled upon as the product name. And then came another question that rocked my world and changed my life.

"Do you mind if we use 'Cooper's' as the brand name? The tea was developed by you, Barry, and you're actually a tea expert. We'd be foolish not to let people know that. You can talk about the product, be its 'face.'"

I agreed. Who wouldn't?

It still had to be tested, though. At this point I had invested a great deal of time and hundreds of thousands of dollars, and there was still no guarantee of success. But success did come. The test markets showed that the public preferred our product to the two national brands. And in 2003, Cooper's Iced Brew Tea was launched across the country. Suddenly I had a national brand. I had to pinch myself—it was an unreal feeling.

Four years later, it is still the preferred tea for 7-Eleven and, with few exceptions, can be found on the fountain in their stores. From there it was pretty easy to expand into restaurants. We offered a superior-tasting tea, a unique story, and a founder who believed in the product enough to put his face on the package. You can't ask for much more than that. In 2006 we sold about forty million glasses of iced tea.

That's a great story in and of itself. But it has another chapter. In 2005, I was back in Dallas to meet with the category manager for my Iced Brew Tea. He was called away on an emergency, and as a courtesy because I had traveled from Colorado, a senior vice president met with me instead. We chatted about our various industries. He explained that the energy drink category was taking off like a rocket. I mentioned that green tea was the hot seller in my business and noted how ironic it was that most people didn't realize that green tea contains natural stimulants. We looked at each other and didn't have to say a thing. A green-tea-based energy drink! This could be big.

Yet again, I was faced with a new challenge. I had learned about fountain beverages. But this would not be a fountain drink. This would be a bottled, ready-to-drink product with a lot of competition. Our research told us that a green energy tea was a concept that people liked. You could get all the stimulation that the current energy drinks offered but through natural ingredients. Once people understood what an energy tea was, they were immediately interested. But that was the crux of the matter. Everyone over the age of five has heard of iced tea; this is not so with energy tea. With my Iced Brew Tea, I "only" had to sell the concept to the management of a restaurant chain. It isn't an easy sell, but only a person or two needs to recognize the high quality and value of the product. The energy tea would involve selling one bottle at time to one person at a time. The difference is huge. One mainly requires good manufacturing practices and excellent customer service. The other requires advertising, promotion, and a constant reinforcement of the product in the marketplace. Companies like Coca-Cola, Pepsi, and Lipton have the financial deep pockets for a huge public launch of their products. They run TV ads, sponsor huge events, pay slotting fees to secure space on the supermarket shelves, and do a host of other things that can cost millions of dollars. Small companies do not have the same resources.

Chapter 31

So launching a ready-to-drink bottled product is an extremely high-risk proposition.

I hired another local Boulder marketing firm, The Creative Alliance, to help tackle the project. They put their best strategic planners and designers to task, and off we went. Being as proud of this product as I am of the others, I was game to put my moniker on this one, too. But knowing that the audience for an energy tea was going to be largely college students and young professionals, I thought we had better test out that idea. In the end, we developed over thirty different designs for our bottle labels looking for the one that best conveyed the sense of adventure, of fearlessness, of never-say-die that I wanted it to project. As it turns out, the image of a sixty-year-old tea master on the front of a bottle apparently did not fill eighteen-year-olds with that feeling. "Who is the old guy?" was one young man's memorable comment.

As the tactful marketing folks told me, everyone liked the fact that a true tea man had developed the product. They also loved hearing about my adventure stories from around the world. They just didn't necessarily want me quite so front and center. We needed to find a new twist on the concept. And a "twist" is exactly what we did. To convey the sense of adventure, we used a safari theme. The front of the bottle features an African thorn tree with a different safari animal for each flavor. But when you twist the bottle around to learn more, you'll find a story from that old guy who has lived the life, developed the tea, and tells the truth.

My favorite part of the packaging, though, is the name—BAZZA High-Energy Tea. Bazza is a nickname that I picked up while traveling in Australia. That was one crazy visit. It could have been worse; I got off easy with "Bazza." Anyhow, I was reminiscing about that trip to a member of my marketing team when suddenly she said, "That's it. That's what we should call the energy tea—BAZZA."

Putting a Face to Cooper Tea

They went to the local college campus and tested it out. It was a hit! At one session, the students actually started chanting "BAZZA, BAZZA, BAZZA." So I thought, *OK. BAZZA it is.*

The verdict is still out on our latest project. Energy tea is a new category with a lot of potential, though. We're developing a cult following of the craziest mix—not only college kids, but Miami DJs who like that it doesn't have a "sugar crash," healthy eaters who like the green tea antioxidants, young women looking for zero-calorie drinks, and rugby players who basically just like that it works and tastes good. If we brought everyone together, we'd have a showstopper of a party.

The beverage business is a tough, competitive world. But while I never joined the armed forces, I *was* raised with their histories. As we battle against the Goliaths for the grocery shelf, the Special Air Service motto, "who dares, wins," seems like a good creed to live by. Hell, the big guys don't even know that I'm considering a trip to Dallas to plan another surprise attack.

Chapter 32
Summing Up

Thomas J. Lipton once owned a tea estate in Ceylon named Dambetenne. He had a cement throne built there, high on the cliffs overlooking his home. The view from this throne is startling. You are perched high in the air, surveying thousands of miles of horizon and sky.

Far below to the right is Lipton's magnificent bungalow, built on a rock outcrop that juts aggressively out into open space. An ornate wooden pagoda perches perilously at the edge. In front of you are fields of endless green rows of tea. The bushes run parallel in the field, but from this viewpoint, they create quilts of green and gold that merge into haze and then into the horizon. The wind is strong up here. It gusts through the rocky crevices and whips the thin strands of grass into moments of frenzy. It is the only sound you hear.

To get to this eagle perch, you must climb a steep cliff behind Lipton's home and traverse a ridge that brings you to his square, cement throne. The rocks are bare, apart from tenacious spots of lichen that mottle the gray stone with autumn colors of green and brown. Lipton must have clambered up with a retinue of porters carrying cushions, drinks, and maybe even a blanket or two; it gets cold very quickly up in the mountains.

I had the opportunity to sit on Lipton's throne once, many years ago. As I sat there, I wondered what Lipton himself had thought while he surveyed his domain. Had he known that he would bring into reality the largest tea company in the world? I suspect not; one never knows where the adventure of tea will lead. I certainly had no idea where tea would take me when I joined the trade.

For over forty years, I have crisscrossed the globe tasting, buying, researching, transporting, blending, and selling this amazing leaf. It has given me a love and respect for tea that endures to this day. It has been the stuff of dreams.

Tea continues to fascinate me, for if you look closely nothing about it is as it seems. These magical leaves have been made into bricks, tablets, and dust. Tea leaves have been whisked to a froth, packed in bags, sold loose, and extracted into solid and liquid concentrates. Tea can be served hot or cold with milk, lemon, sugar, or spices. Tea is consumed in every country, and it remains the world's most popular beverage after water. Tea has opened up foreign lands to development, and it even played a role in starting a war—remember the chests of tea being thrown into Boston Harbor? It has been the drink of kings and emperors and slakes the thirst of peasants as well. From one bush come thousands of styles of tea. And now evidence is mounting that the antioxidants in tea can be very good for you as well.

To offer a cup of tea is a courteous, polite act, and in accepting, one joins in a civilized exchange. It is hard to be rude sharing a pot of tea. In the end, perhaps, that is what appeals to me most about tea. In a world full of torment, turmoil, trouble, and strife, tea can be an oasis. Tea is also history; it is romance; it is the stuff of dreams, and yet it is common to almost all societies.

Chapter 32

The beautiful Lipton tea estate in Ceylon.

You may think I am overstating the importance of tea or putting it on too high a pedestal. You may be right, but think how far tea must travel and how many people have worked hard and fought to establish the gardens that grow those two leaves and a bud. Be grateful for their efforts. Hold your next cup close and breathe deep the fragrance. Savor it, and enjoy your moment of tranquility. Life is a series of moments; tea will make this a good one.